W9-COF-178

thomson•com

changing the way the world learns℠

To get extra value from this book for no additional cost, go to:

http://www.thomson.com/wadsworth.html

thomson.com is the World Wide Web site for Wadsworth/ITP and is your direct source to dozens of on-line resources. *thomson.com* helps you find out about supplements, experiment with demonstration software, search for a job, and send e-mail to many of our authors. You can even preview new publications and exciting new technologies.

thomson.com: *It's where you'll find us in the future.*

PHOTO ATLAS

FOR BOTANY

James W. Perry
University of Wisconsin–Fox Valley

David Morton
Frostburg State University

WADSWORTH PUBLISHING COMPANY

I T P ®   An International Thomson Publishing Company

Belmont, CA • Albany, NY • Bonn • Boston • Cincinnati • Detroit • London • Madrid • Melbourne
Mexico City • New York • Paris • Singapore • Tokyo • Toronto • Washington

Biology Editor	JACK CAREY
Assistant Editor	MICHAEL BURGREEN
Editorial Assistant	KERRI ABDINOOR
Production Service	CAROL CARREON LOMBARDI
Managing Designer	CAROLYN DEACY
Print Buyer	KAREN HUNT
Copy Editor	MARY ROYBAL
Cover Designer	GARY HEAD
Cover Photographs	JAMES W. PERRY
Composition and Art Preparation	FOG PRESS
Printer	BANTA

FOR MORE INFORMATION, CONTACT:

Wadsworth Publishing Company
10 Davis Drive
Belmont, California 94002
USA

International Thomson Publishing Europe
Berkshire House 168-173
High Holborn
London, WC1V 7AA

Thomas Nelson Australia
102 Dodds Street
South Melbourne 3205
Victoria, Australia

Nelson Canada
1120 Birchmount Road
Scarborough, Ontario
Canada M1K 5G4

International Thomson Editores
Campos Eliseos 385, Piso 7
Col. Polanco
11560 México D.F. México

International Thomson Publishing GmbH
Königswinterer Strasse 418
53227 Bonn
Germany

International Thomson Publishing Asia
221 Henderson Road
#05-10 Henderson Building
Singapore 0315

International Thomson Publishing Japan
Hirakawacho Kyowa Building, 3F
2-2-1 Hirakawacho
Chiyoda-ku, Tokyo 102
Japan

1 2 3 4 5 6 7 8 9 10

ISBN 0-534-52938-0

DEDICATED TO

JOY B. PERRY

LIFE PARTNER, SPOUSE, BEST FRIEND,

SHARER OF LOVE FOR THE LIVING WORLD

AND TO

RAY F. EVERT

PLANT ANATOMIST, MENTOR, MAJOR PROFESSOR, AND FRIEND

Brief Contents

DETAILED CONTENTS

PREFACE

Photo Atlas for Botany is intended for use in both introductory and advanced plant biology courses. Its subjects will serve as a useful reference in both pure and applied plant sciences, including general botany, plant taxonomy, plant anatomy, agronomy, and plant pathology.

Following on the success of and enthusiasm for *Photo Atlas for Biology* and *Photo Atlas for Anatomy and Physiology*—and motivated by the desire for a more comprehensive treatment of plant biology—comes this compendium of botanical images. We began producing color photo atlases because of their clear advantages of usefulness and accuracy—not only in the laboratory but as review materials to facilitate students' long-term memory. As in all biology courses, the laboratories in botany are visual experiences, full of images of living specimens and microscopic preparations. This atlas will allow students a comprehensive, ongoing review of specimens at any time or place.

The *Photo Atlas for Botany* will be a useful supplement to lab manuals, whether commercial or in-house: Most commercial manuals still contain only limited color materials, and although instructor-created manuals are tailored to a specific course, they are often poorly illustrated. This book includes a central core of images common to all botany courses and many others that may help instructors expand their course offerings.

The real utility of any photo atlas is the opportunity it gives students to review material when "the real thing" is not immediately available. Although CD-ROMs are becoming more common, it's hard to curl up with a computer—and despite our most fervent wishes, paper copy remains the budget-conscious choice of most students. Even in the laboratory with instructor present, students past and present know all too well the long wait for confirmation that what they're looking at is the "right thing." This atlas will serve students in both of these regards.

A book is conceived and born through the collaborative efforts of the authors and many other creative people. Special thanks to Jack Carey, our publisher at Wadsworth, who continues to guide us with his vision; to the Wadsworth in-house production and marketing team; and to Fog Press and Carol Lombardi.

We especially wish to thank the several contributors who provided images not in our collection. Among those are John and Joyce Limbach, Biodisc, Inc., 6963 Easton Court, Sarasota, FL 34238; and Paul Conant, Triarch Microslides, Inc., P.O. Box 98, Ripon, WI 54971 for allowing us to use several one-of-a kind images. An interactive CD-ROM and laserdisc containing over 4,700 microslide images are available from Biodisc. And finally, special thanks to Ray Everta, Susan Eichhorn, and Claudia Lipke of the University of Wisconsin Department of Botany.

ABOUT THE AUTHORS

JIM PERRY is the Campus Executive Officer, Dean, and Professor of Biological Sciences at the University of Wisconsin—Fox Valley. Although his primary functions are now administrative, he returns to the classroom each spring to teach General Botany, which he calls his "inner sanctum and reminder of what this is all

James W. Perry   *David Morton*

about." He holds a B.S. in Zoology and Secondary Education, M.S. in Botany and Zoology, and Ph.D. in Botany and Plant Pathology, all from the University of Wisconsin at Madison. Prior to returning to Wisconsin, he was a faculty member at Frostburg State University, serving as Chair of Biology. His teaching experience includes introductory Biology courses and upper-level offerings in fungi, algae, the plant kingdom, and electron microscopy. Most recently, he has participated in courses assisting graduate students to learn the art and science of teaching biology.

DAVE MORTON is Chair of the Biology Department at Frostburg State University. After earning a B.S. in Zoology and teaching junior high school, he attended Cornell University, where he received a Ph.D. with a major in Histology and minors in Physiology and Biochemistry. For more than twenty years he has taught numerous introductory, upper-level, and graduate biology courses at the college/university level. Some of his more interesting research publications describe aspects of iron and fluid balance in vampire bats.

We welcome your comments about the balance we've achieved and hope you will use the addresses below to let us know how we can help you further. Our most rewarding moments occur when students call or e-mail us about our efforts. Questions concerning this text should be directed to:

James W. Perry
University of Wisconsin–Fox Valley
1478 Midway Road, P.O. Box 8002
Menasha, WI 54952-8002
phone: 920/832-2610
e-mail: jperry@uwc.edu

David Morton
Department of Biology
Frostburg State University
Frostburg, MD 21532-1099
phone: 301/687-4355
e-mail: d_morton@fre.fsu.umd.edu

Let's examine two images and their accompanying legends.

Figure 96a   Lily flower (Turk's cap lily, *Lilium superbum; Liliaceae*). Sections of floral buds (unexpanded flowers) are used to create slides depicted in Figures 96b through 97b and Figures 98a through 99e (live, 0.25×). (Photo by J. W. Perry)

Figure 98a   Lily ovary (prep. slide, c.s., 8×). (Photo by J. W. Perry)

"Lily flower," in boldface type, describes the subject of the photo. You will find most boldfaced words in the Subject Index that begins on page 135.

"Turk's cap lily" is the common name of the organism.

Lilium superbum is the scientific binomial: *Lilium* the genus, *superbum* the specific epithet. Together, the genus and the specific epithet define the species of this plant. You will find genus names in the Genera Index that begins on page 131.

Liliaceae is the family to which this plant belongs. Family names have been included only with images of flowering plants.

"live" means that the photo was taken of a living specimen.

"0.25×" means that the image is 1/4 of actual size.

"Lily ovary" is the subject of this image.

"prep. slide" means that the image was created using a light microscope and a commercially prepared glass microscope slide.

"c.s." is the abbreviation for cross section, meaning that the ovary has been sectioned transversely. Typically, a commercial microscope slide would have this abbreviation on its label. (See also other abbreviations listed on page xi.)

"8×" means that the image on the page is eight times the actual size of the specimen.

Magnifications

We chose to calculate magnifications of the images in *Photo Atlas for Botany* as compared to the size of the actual specimen, rather than indicate only the microscope objective lens or magnification that was used to capture the original image. Thus any magnification of 1× or greater means that the image on the page is larger than the specimen is in life. Conversely, any magnification less than 1 means the printed image is smaller than life-size.

You can do a fairly accurate calculation of the specimen's actual size by measuring the printed image:

Figure 96a is 31 mm across. 0.25 = 1/4. 31 mm × 4 = 124 mm. This flower is 124 mm across in life.

Similarly, the diameter of the lily ovary in Figure 98a measures 48 mm. 48 mm / 8 = 6 mm. In life, this ovary was 6 mm across.

Taxonomy

Taxonomic classification is a source of never-ending conjecture. Recently, the botanical category known as "division" was officially replaced by "phylum," bringing into agreement this level of classification for all organisms. Beyond that major change, however, a gulf of discord remains. If you review ten botany books, you are sure to find a least nine different schemes. For example, most algae are being placed in the phylum Protista, and even here there are those who believe "Protista" should be "Protoctista." The classification used herein reflects the most current thinking among many plant taxonomists and systematists. We recognize that each instructor may have her or his own preferences.

Abbreviations

c.s.	cross section
d.i.c.	differential interference microscopy
l.s.	longitudinal section
live	photo taken of a living specimen
prep. slide	photo taken of a prepared microscope slide
sec.	section
s.e.m.	scanning electron microscopy
t.e.m.	transmission electron microscopy
t.s.	tangential section
w.m.	whole mount

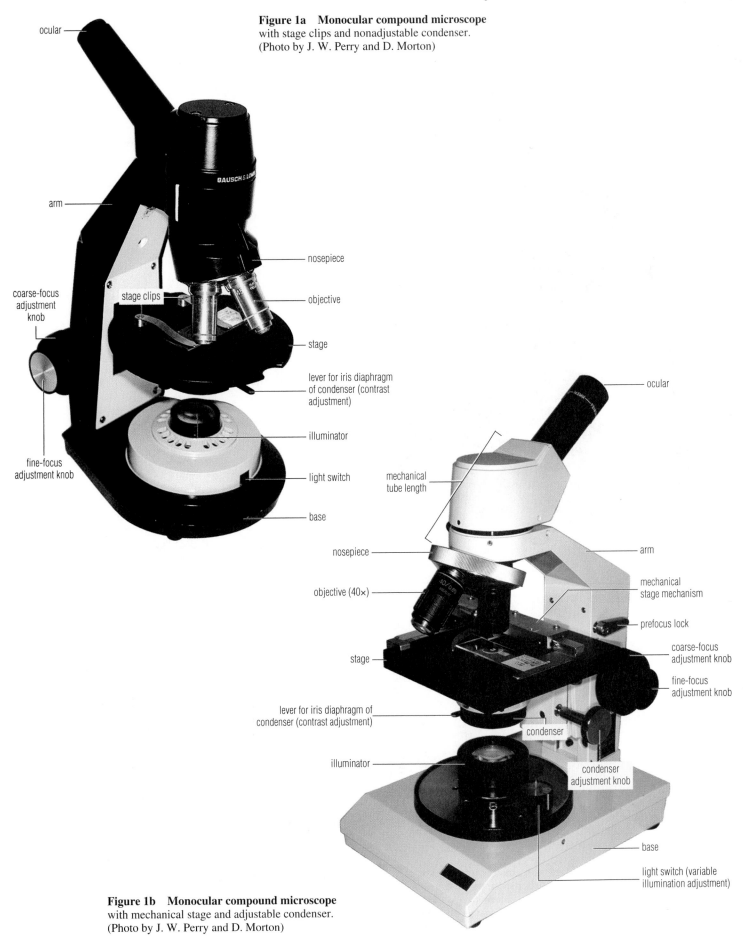

Figure 1a   Monocular compound microscope
with stage clips and nonadjustable condenser.
(Photo by J. W. Perry and D. Morton)

ocular

arm

nosepiece

objective

coarse-focus adjustment knob

stage clips

stage

lever for iris diaphragm of condenser (contrast adjustment)

illuminator

light switch

fine-focus adjustment knob

base

ocular

mechanical tube length

nosepiece

arm

objective (40×)

mechanical stage mechanism

prefocus lock

coarse-focus adjustment knob

stage

fine-focus adjustment knob

lever for iris diaphragm of condenser (contrast adjustment)

condenser

illuminator

condenser adjustment knob

base

light switch (variable illumination adjustment)

Figure 1b   Monocular compound microscope
with mechanical stage and adjustable condenser.
(Photo by J. W. Perry and D. Morton)

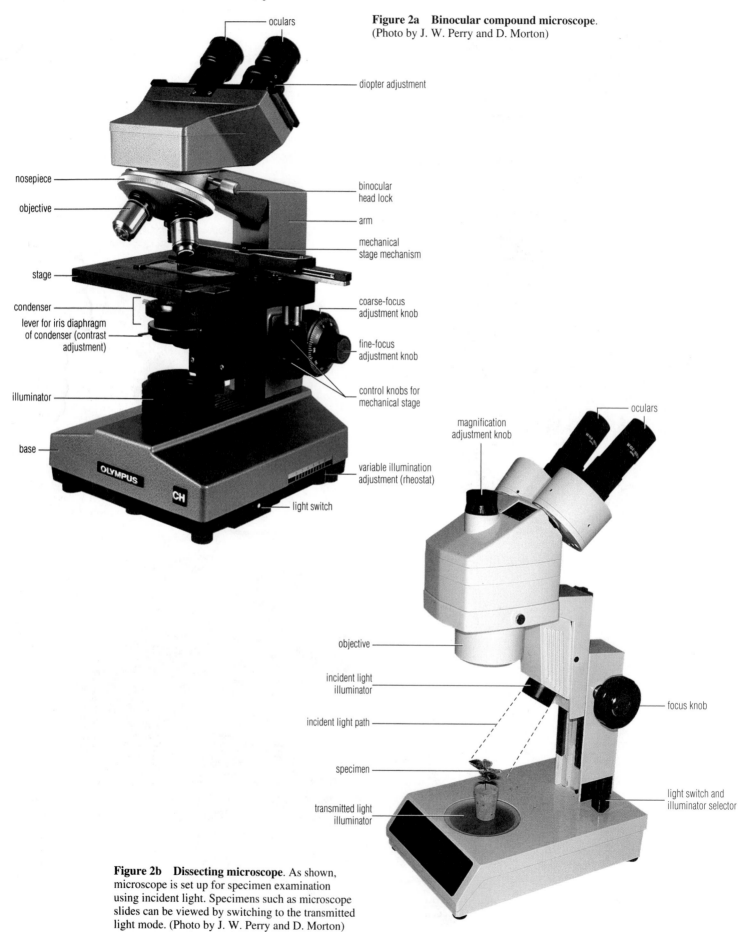

oculars

Figure 2a   Binocular compound microscope.
(Photo by J. W. Perry and D. Morton)

diopter adjustment

nosepiece

binocular
head lock

objective

arm

mechanical
stage mechanism

stage

coarse-focus
adjustment knob

condenser

lever for iris diaphragm
of condenser (contrast
adjustment)

fine-focus
adjustment knob

control knobs for
mechanical stage

illuminator

base

variable illumination
adjustment (rheostat)

light switch

oculars

magnification
adjustment knob

objective

incident light
illuminator

focus knob

incident light path

specimen

light switch and
illuminator selector

transmitted light
illuminator

Figure 2b   Dissecting microscope. As shown,
microscope is set up for specimen examination
using incident light. Specimens such as microscope
slides can be viewed by switching to the transmitted
light mode. (Photo by J. W. Perry and D. Morton)

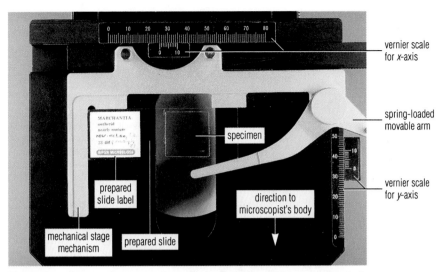

Figure 3a Correct placement of prepared microscope slide on **mechanical stage**. The slide label is oriented so that it can be read by the microscopist. (Photo by J. W. Perry)

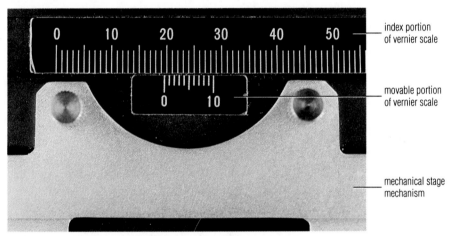

Figure 3b **Vernier scale** on mechanical stage of compound microscope. The correct reading is 19.6 mm. (Photo by J. W. Perry)

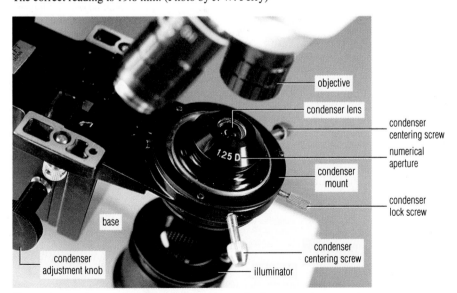

Figure 3c Adjustable **condenser** on a compound microscope (stage removed). The engraved number is the condenser lens's numerical aperture. The "D" following the numerical aperture indicates that the condenser lens is to be used dry—that is, with no oil atop the lens to make contact with the bottom of the slide. (Photo by J. W. Perry)

Figure 3d **Ocular** (eyepiece) removed from microscope. The markings on the ocular indicate this to be a "wide field" lens with a magnification of 10 (10×) and a focal length of 18.5 mm. (Photo by J. W. Perry)

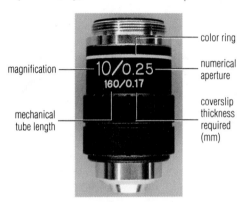

Figure 3e **10× objective**. Engravings indicate that the numerical aperture is 0.25, the mechanical tube length must be 160 mm, and a coverslip 0.17 mm thick (No. 2 coverslip) must be used over the specimen for optimal resolution. See Figure 1b for illustration of mechanical tube length. Color rings of objectives indicate their magnification but may vary by manufacturer. (Photo by J. W. Perry)

Figure 3f **100× oil immersion objective**. Engravings indicate that the numerical aperture is 1.25, the mechanical tube length must be 160 mm, and the objective is to be used with a slide having a coverslip 0.17 mm thick (No. 2 coverslip). The black ring indicates that the gap between the lens and the microscope slide's coverslip is to be filled with immersion oil. (Photo by J. W. Perry)

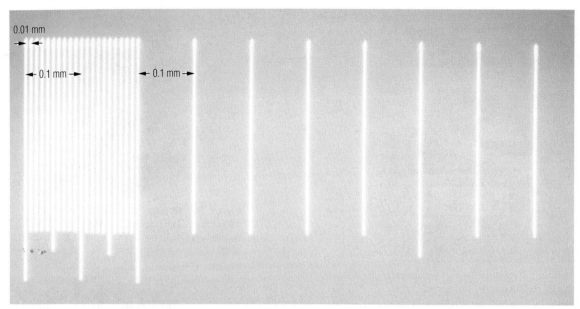

Figure 4a **Stage micrometer** as viewed through a microscope. Stage micrometers are used to calibrate ocular micrometers for future measurements of microscopic objects being viewed. The distance between the lines that are closest together (left side of this image) is 0.01 mm (=10 µm), and the distance between the lines that are farther apart is 0.1 mm (=100 µm) (160×). (Photo by J. W. Perry)

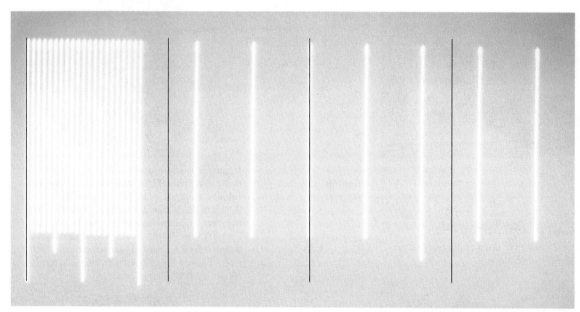

Figure 4b **Stage micrometer** (white lines) with **ocular micrometer** (black lines) superimposed. As illustrated, each ocular micrometer division equals 0.25 mm (250 µm). Here's how that was determined: The leftmost black ocular line is placed over the zero-line (white) stage micrometer line. Count ocular lines until you find one that matches *exactly* with a stage micrometer line. In this case, the second ocular division line from the zero line falls on the 0.5-mm stage micrometer line, and 0.5 mm/2 lines = 0.25 mm/ocular division. *The ocular divisions must be calculated for each objective and then recorded for future reference* (160×). (Photo by J. W. Perry)

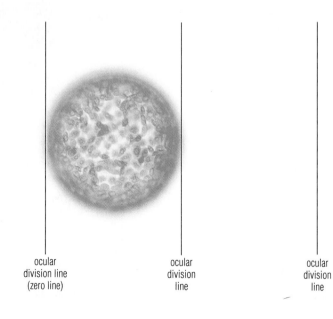

Figure 4c **Size determination** of an object with an **ocular micrometer**. Using the information determined in Figure 4b, we judge the size of this spherical algal cell to be 0.25 mm (250 µm) in diameter. Note that if the ocular line did not fall precisely on the right edge of the object, some interpolation would be needed (160×). (Photo by J. W. Perry)

ocular division line (zero line)

ocular division line

ocular division line

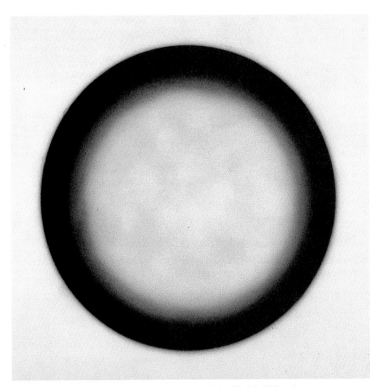

Figure 5a **Free air bubble**, a common "what's this?" in wet mounts (420×). (Photo by J. W. Perry)

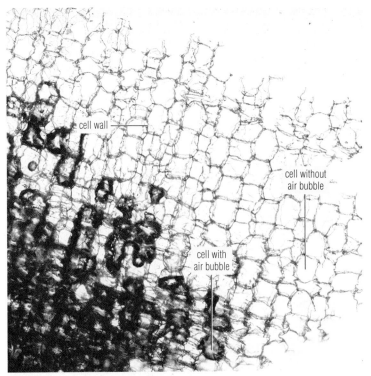

Figure 5b **Cork cells**. The darkened areas are cells filled with air (c.s., 160×). (Photo by J. W. Perry)

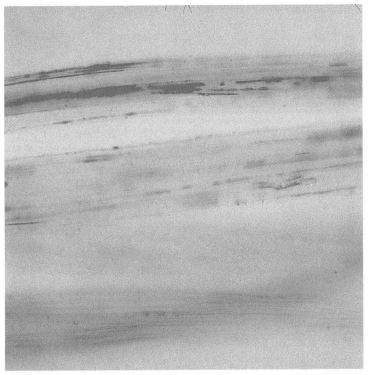

Figure 5c **Low-contrast** photo of unstained cotton fibers taken through a microscope with iris diaphragm fully open (prep. slide, w.m., 140×). (Photo by J. W. Perry)

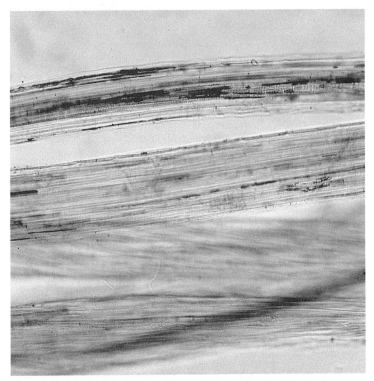

Figure 5d **Increased-contrast** photo of unstained cotton fibers taken through a microscope with iris diaphragm fully closed (prep. slide, w.m., 140×). (Photo by J. W. Perry)

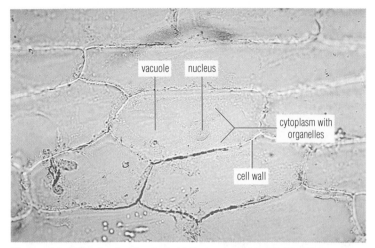

Figure 6a   **Unstained** onion bulb (*Allium cepa*) leaf cells. **Brightfield** ("normal") **microscopy** (live, w.m., 225×). (Photo by J. W. Perry)

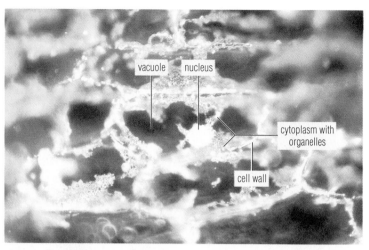

Figure 6b   Onion bulb (*Allium cepa*) leaf cells viewed with **darkfield microscopy** (live, w.m., 225×). (Photo by J. W. Perry)

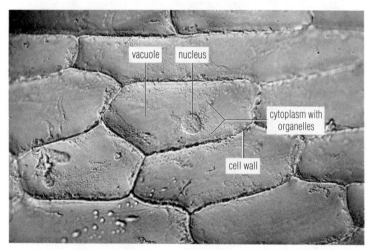

Figure 6c   Onion bulb (*Allium cepa*) leaf cells viewed with **differential interference contrast (DIC = Normarski) microscopy**. This special technique produces a three-dimensional impression of the cells' organelles. Student microscopes are rarely equipped to produce such images (live, w.m., 225×). (Photo by J. W. Perry)

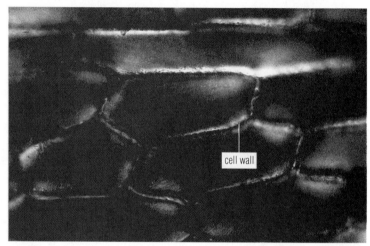

Figure 6d   Onion bulb (*Allium cepa*) leaf cells viewed with **polarizing microscopy**. Crystalline structures glow in a dark background with this technique. The cell walls of these cells have a slight crystallinity (live, w.m., 225×). (Photo by J. W. Perry)

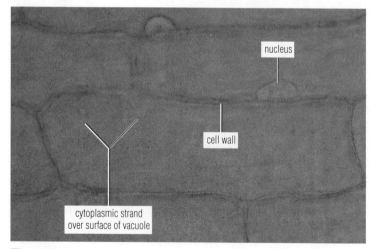

Figure 6e   Onion bulb (*Allium cepa*) leaf cells **stained with safranin**. Brightfield microscopy (live, w.m., 225×). (Photo by J. W. Perry)

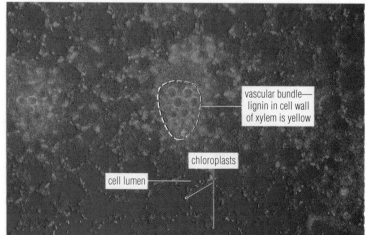

Figure 6f   Begonia (*Begonia* sp.) leaf cells viewed with **fluorescence microscopy**. Certain plant compounds autofluoresce when illuminated with ultraviolet light. Here the lignin in some of the cell walls fluoresces yellow, and the chlorophyll fluoresces red. Stains called *fluorochromes* can be added to cause other substances to fluoresce (live, w.m., 60×). (Photo by J. W. Perry)

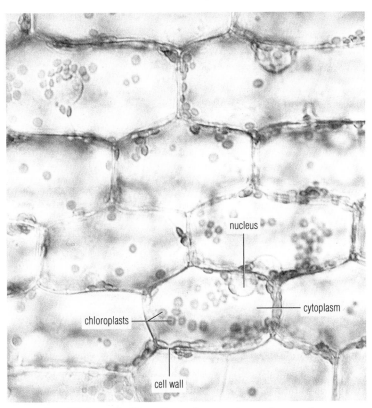

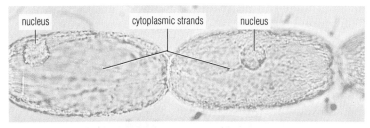

Figure 7c *Tradescantia* **stamen hair cells**. The vacuole of these cells contains a pink anthocyanin pigment. The cytoplasm is restricted to thin strands that overlay the vacuole (live, w.m., 375×). (Photo by J. W. Perry)

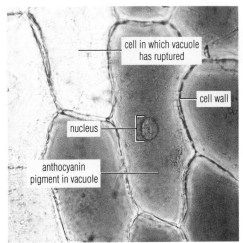

Figure 7d Red onion bulb (*Allium cepa*) leaf **parenchyma cells**. In some cells, the vacuole is filled with a red **anthocyanin pigment**; in others, it has ruptured, releasing the pigment (live, w.m., 230×). (Photo by J. W. Perry)

Figure 7a *Elodea* **cells**. The cell membrane and vacuolar membrane are too thin to be resolved with the light microscope. Mounted in an isotonic solution, these cells are **turgid** (live, w.m., 600×). (Photo by J. W. Perry)

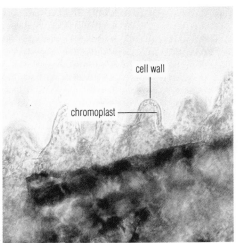

Figure 7e Flower petal **parenchyma cells** of primula (*Primula kewensis*) containing yellow **chromoplasts** (live, w.m., 375×). (Photo by J. W. Perry)

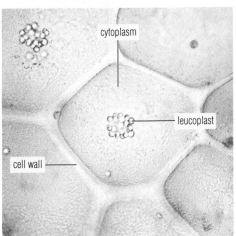

Figure 7f Leaf cells of *Zebrina* with **leucoplasts** clustered around the nucleus. The faint coloration is due to anthocyanin pigment in each cell's vacuole (live, w.m., 375×). (Photo by J. W. Perry)

Figure 7b *Elodea* **cells** that have been **plasmolyzed** by placing them in a hypertonic solution (live, w.m., 600×). (Photo by J. W. Perry)

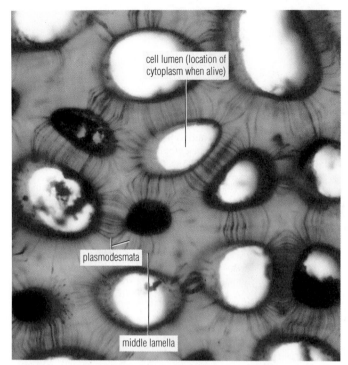

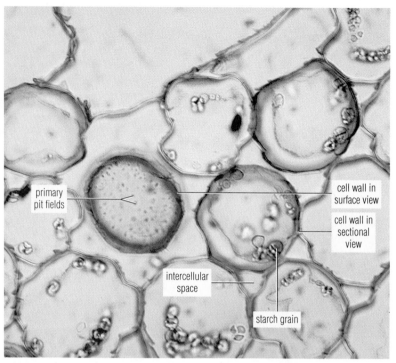

Figure 8a　**Plasmodesmata** in thick walls of parenchyma cells from the endosperm of *Diospyros* (persimmon). The **middle lamella** cementing together adjacent cells is perpendicular to the fine plasmodesmata (prep. slide, c.s., 520×). (Photo by J. W. Perry)

Figure 8b　**Primary pit fields** (thin areas in the primary cell wall) as they appear when viewing the surface of the cell wall in parenchyma cells (*Ranunculus* root). Primary pit fields look like holes (even though they are not) because more light passes through the thin area (prep. slide, c.s., 325×). (Photo by J. W. Perry)

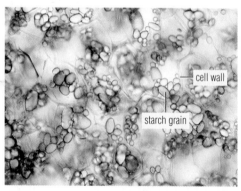

Figure 8c　**Starch grains** in parenchyma cells of a potato (*Solanum tuberosum*) tuber (live, sec., 90×). (Photo by J. W. Perry)

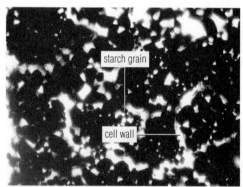

Figure 8d　**Starch grains** in potato tuber stained with iodine (I_2KI) solution (live, sec., 90×). (Photo by J. W. Perry)

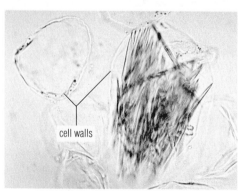

Figure 8e　**Raphides** crystals from macerated leaf cells of bowstring hemp (*Sansevieria*), unstained, brightfield microscopy (w.m., 300×). (Photo by J. W. Perry)

Figure 8f　**Raphides** crystals from macerated leaf cells of bowstring hemp (*Sansevieria*), unstained, polarized light microscopy. Crystals appear bright in a dark background with polarized light (w.m., 300×). (Photo by J. W. Perry)

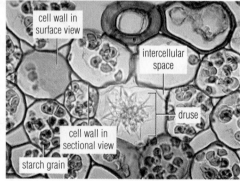

Figure 8g　**Druse crystal** in stem of wax plant (*Hoya carnosa*) (prep. slide, c.s., 375×). (Photo by J. W. Perry)

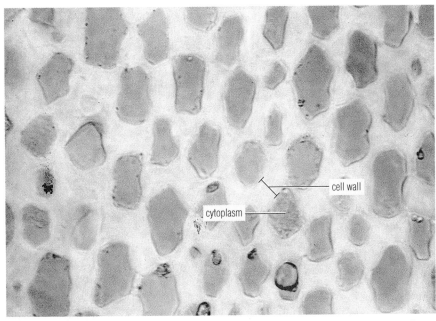

Figure 9a Unstained **collenchyma cells** from the petiole of celery (*Apium*). Darkened areas are the cell lumens (regions where cytoplasm is located), white areas the cell walls (live, c.s., 670×). (Photo by J. W. Perry)

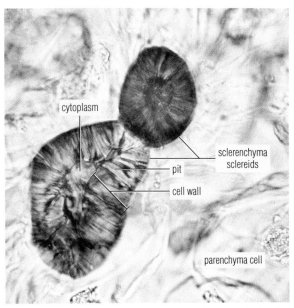

Figure 9b **Sclerenchyma sclereid cells** from the flesh of pear (*Pyrus*) fruit, stained with phloroglucinol in 20% hydrochloric acid. Lines in cell walls are pits (channels) in the secondary cell wall (live, w.m., 450×). (Photo by J. W. Perry)

Figure 9c **Sclerenchyma fibers** and **raphides** from macerated basswood (*Tilia*) bark, as seen with polarized light (live, w.m., 530×). (Photo by J. W. Perry)

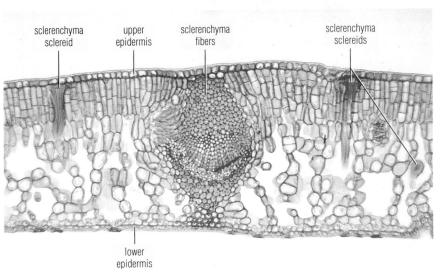

Figure 9d **Sclerenchyma fibers and sclereids** in a leaf (*Osmanthus*). Sclereids run from epidermis to epidermis, providing rigidity. Fibers surround the vascular strand (prep. slide, c.s., 520×). (Photo by J. W. Perry)

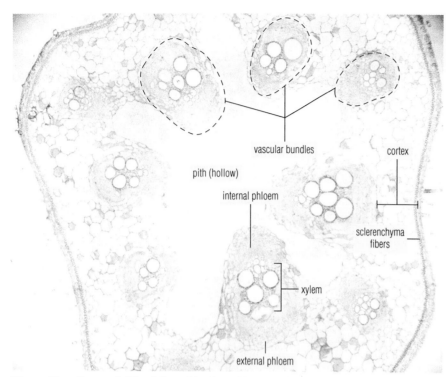

vascular bundles

pith (hollow)

internal phloem

cortex

sclerenchyma fibers

xylem

external phloem

Figure 10a Prepared slide of **Cucurbita** (squash) stem. Figures 10b–11b are higher magnifications of areas of this stem (prep slide, c.s., 17×). (Photo by J. W. Perry)

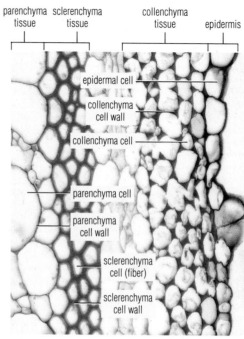

parenchyma tissue   sclerenchyma tissue   collenchyma tissue   epidermis

epidermal cell

collenchyma cell wall

collenchyma cell

parenchyma cell

parenchyma cell wall

sclerenchyma cell (fiber)

sclerenchyma cell wall

Figure 10b Edge of squash (*Cucurbita*) stem showing all three plant cell types: **parenchyma**, **collenchyma**, and **sclerenchyma** (prep. slide, c.s., 230×). (Photo by J. W. Perry)

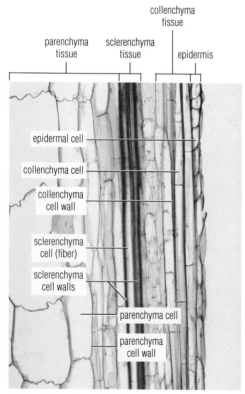

collenchyma tissue

parenchyma tissue   sclerenchyma tissue   epidermis

epidermal cell

collenchyma cell

collenchyma cell wall

sclerenchyma cell (fiber)

sclerenchyma cell walls

parenchyma cell

parenchyma cell wall

Figure 10c Longitudinal section of edge of squash (*Cucurbita*) stem showing three plant cell types: **parenchyma**, **collenchyma**, and **sclerenchyma** (prep. slide, l.s., 90×). (Photo by J. W. Perry)

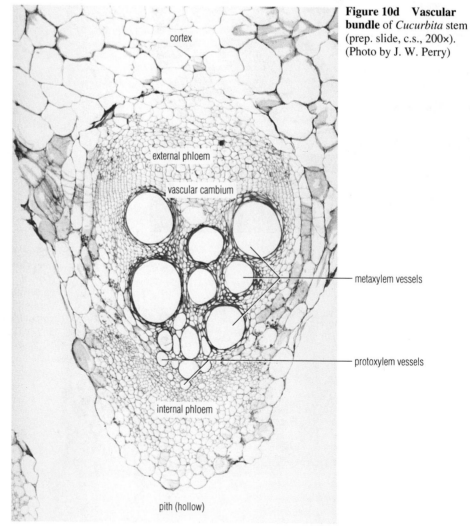

cortex

external phloem

vascular cambium

internal phloem

pith (hollow)

metaxylem vessels

protoxylem vessels

Figure 10d **Vascular bundle** of *Cucurbita* stem (prep. slide, c.s., 200×). (Photo by J. W. Perry)

phloem tissue   xylem tissue

Figure 11a **Xylem** and **phloem** tissue in squash (*Cucurbita*) stem in longitudinal section (prep. slide, l.s., 245×). (Photo by J. W. Perry)

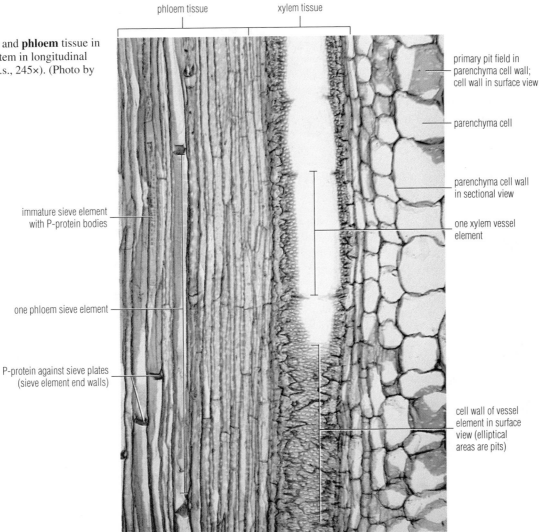

primary pit field in parenchyma cell wall; cell wall in surface view

parenchyma cell

immature sieve element with P-protein bodies

parenchyma cell wall in sectional view

one xylem vessel element

one phloem sieve element

P-protein against sieve plates (sieve element end walls)

cell wall of vessel element in surface view (elliptical areas are pits)

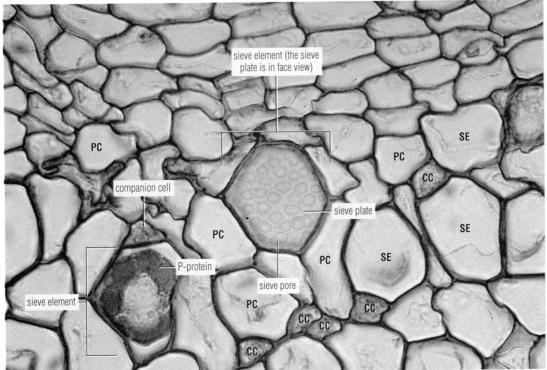

Figure 11b **Phloem** of squash (*Cucurbita*) in cross section. Sieve elements (specifically, sieve tube members) are labeled SE, companion cells CC, parenchyma cells PC. P-protein in this preparation is stained gray-green, in contrast to that stained red in Figure 11a (prep. slide, c.s., 860×). (Photo by J. W. Perry)

sieve element (the sieve plate is in face view)

PC

PC

SE

PC

CC

companion cell

sieve plate

SE

P-protein

PC

SE

sieve element

PC

sieve pore

PC

CC

CC

CC

CC

CC

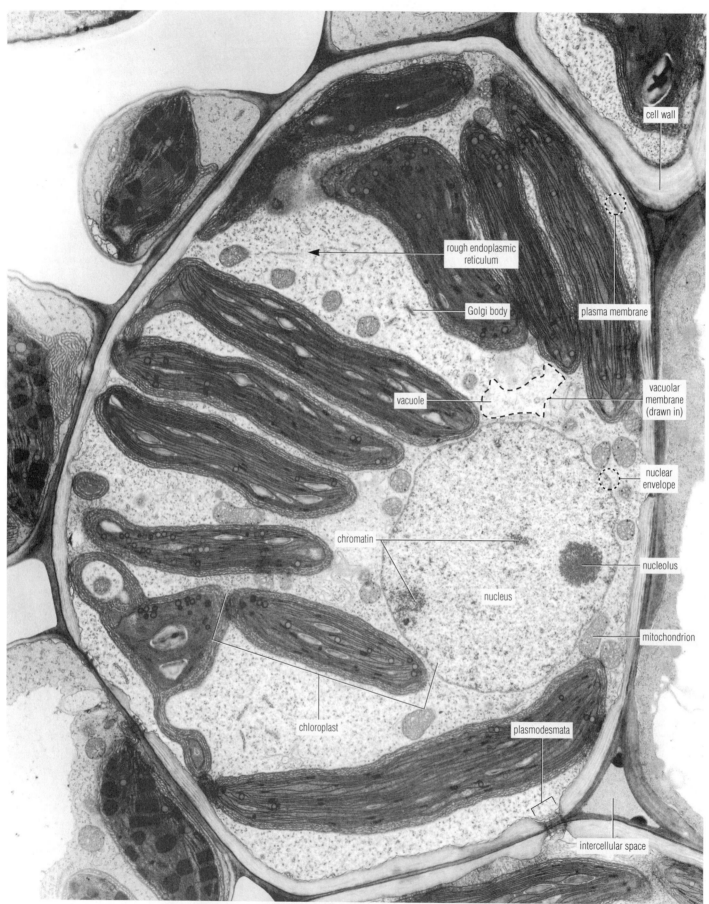

Figure 12a   Transmission electron micrograph (t.e.m.) of a **leaf cell** of corn (*Zea mays*) (t.e.m., c.s., 3450×). (Photo courtesy R. F. Evert and M. A. Walsh)

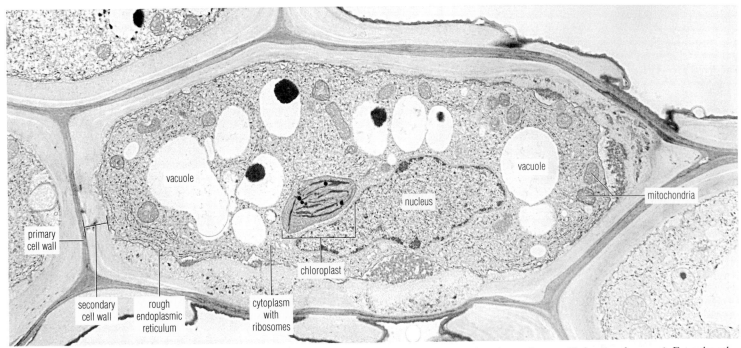

Figure 13a   Transmission electron micrograph of a **stem cell** with a thick **secondary wall** from xylem tissue of potato (*Solanum tuberosum*). Even though the cell is located deep within the stem, it still has chloroplasts, one of which is visible here (t.e.m., c.s. 4000×). (Photo by J. W. Perry)

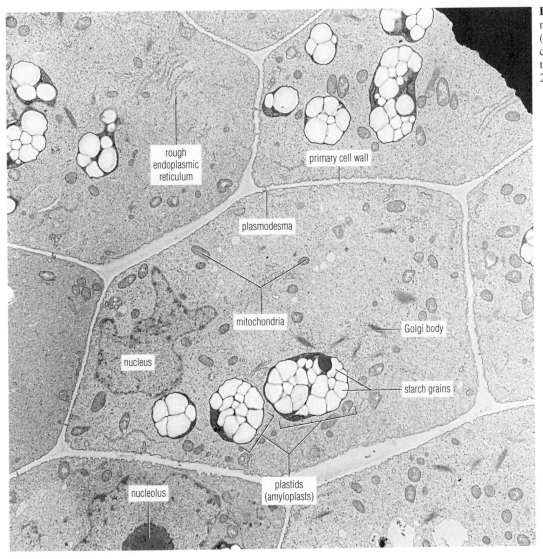

Figure 13b   Transmission electron micrograph of a **root cap cell** of potato (*Solanum tuberosum*). The large, starch-containing plastids are active in gravitropic responses of the root (t.e.m., c.s., 2000×). (Photo by J. W. Perry)

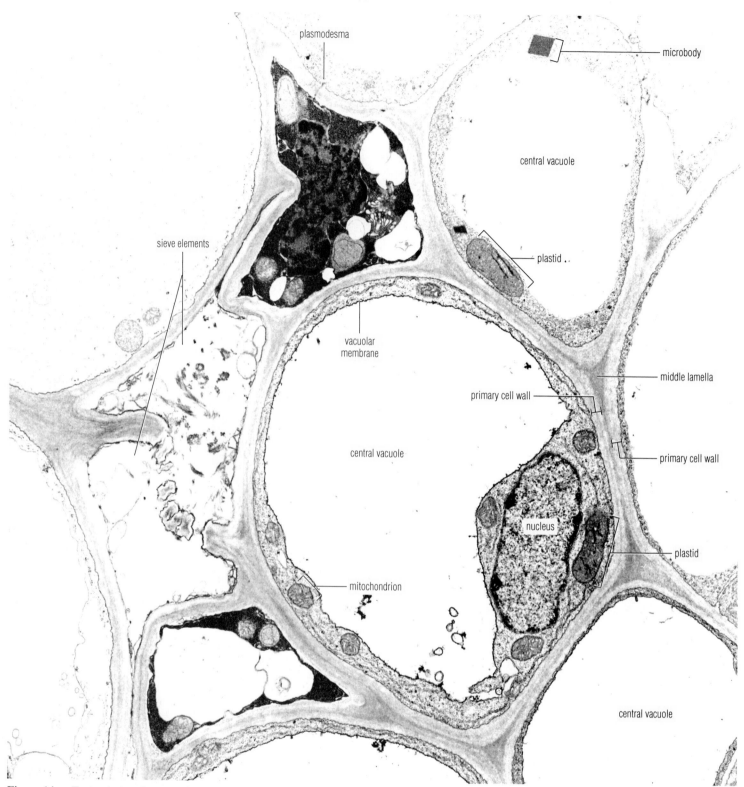

plasmodesma

microbody

central vacuole

sieve elements

plastid

vacuolar membrane

middle lamella

primary cell wall

central vacuole

primary cell wall

nucleus

plastid

mitochondrion

central vacuole

Figure 14a   Transmission electron micrograph of cells within the **phloem** tissue of a potato stem. Note the diversity of cell structure. Some cells have a very large central **vacuole**, often said to be "characteristic" of plant cells, but as shown in Figures 12a, 13a and 13b, that is not always the case (t.e.m., c.s., 8500×). (Photo by J. W. Perry)

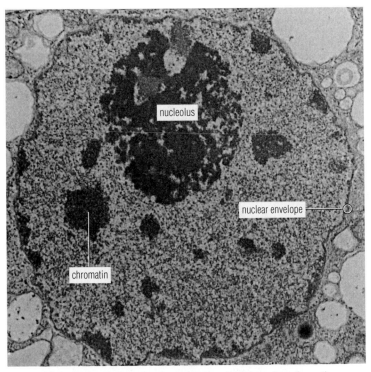

Figure 15a Transmission electron micrograph of a **nucleus** from the whisk fern *Psilotum nudum* (t.e.m., c.s., 9000×). (Photo by J. W. Perry)

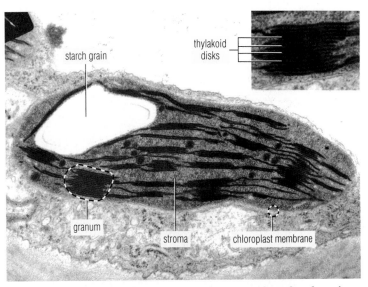

Figure 15b Transmission electron micrograph of a **chloroplast** from the horsetail *Equisetum hymale* (t.e.m., 6000×). Inset: a single **granum** (t.e.m., l.s., 15,000×). (Photo by R. R. Dute)

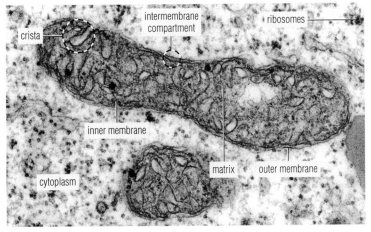

Figure 15c Transmission electron micrograph of a **mitochondrion** from a cell of the fern *Regnellidium*. The intermembrane compartment is the region between the inner and outer membranes (t.e.m., sec., 13,500×). (Photo by S. E. Eichhorn)

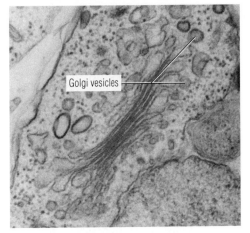

Figure 15d Transmission electron micrograph of a **Golgi body**, sometimes also called a **dictyosome**, in a plant cell. See Figure 16a for the appearance of a Golgi body in face view (t.e.m., l.s., 17,000×). (Photo by W. A. Jensen)

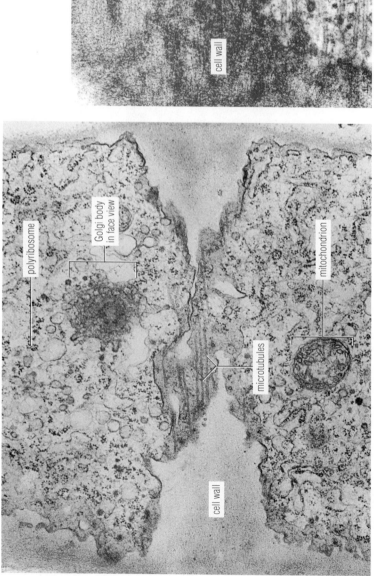

Figure 16a   Transmission electron micrograph of **microtubules** in longitudinal section running just beneath the cell wall of a potato (*Solanum tuberosum*) stem cell. Note also **polyribosomes** and **Golgi body** in face view (t.e.m., sec., 10,000×). (Photo by J. W. Perry)

Figure 16b   Higher-magnification transmission electron micrograph of **microtubules** in longitudinal section (t.e.m., l.s., 70,000×). (Photo by E. H. Newcomb.)

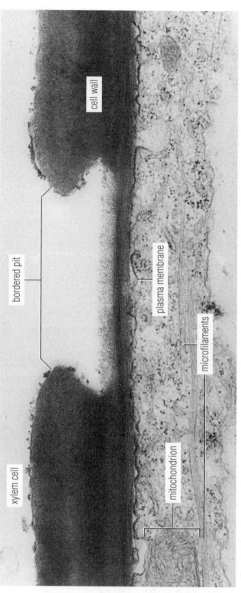

Figure 16d   Transmission electron micrograph of **microfilaments** from a potato stem cell. The adjacent xylem cell has a **bordered pit** in its cell wall (t.e.m.,c.s., 16,000×). (Photo by J. W. Perry)

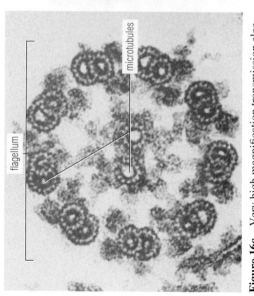

Figure 16c   Very high magnification transmission electron micrograph of **microtubules** of a flagellum in cross section (t.e.m., c.s., 542,000×). (Photo courtesy K. Fujiwara, from *J. Cell Biol.* 59:267 [1963], by permission of the Rockefeller University Press)

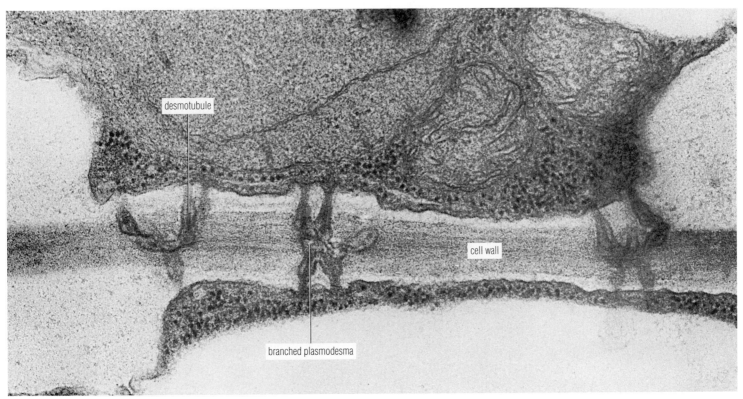

Figure 17a　Transmission electron micrograph of **plasmodesmata** traversing the cell wall of a potato stem cell as seen in sectional view (t.e.m., sec., 30,000×). (Photo by J. W. Perry)

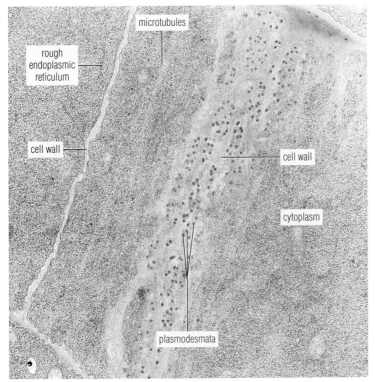

Figure 17b　Transmission electron micrograph of **plasmodesmata** in the cell wall of a potato stem cell as seen in face (surface) view. The granularity in the cytoplasm is due to an abundance of **ribosomes** (t.e.m., sec., 10,000×). (Photo by J. W. Perry)

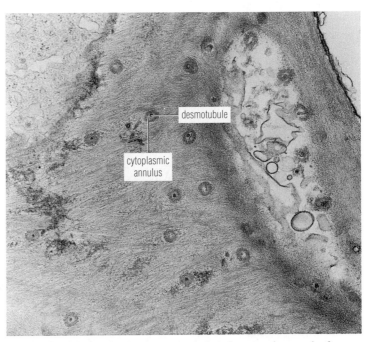

Figure 17c　High-magnification transmission electron micrograph of **plasmodesmata** in the cell wall of a potato stem cell as seen in face (surface) view (t.e.m., sec., 22,000×). (Photo by J. W. Perry)

Figure 18a **Onion** (*Allium cepa*) **root tip**. Cells illustrated in Figures 18b–19e are found in this region of the root. All figures are from prepared slides (l.s., 140×). (Photo by J. W. Perry)

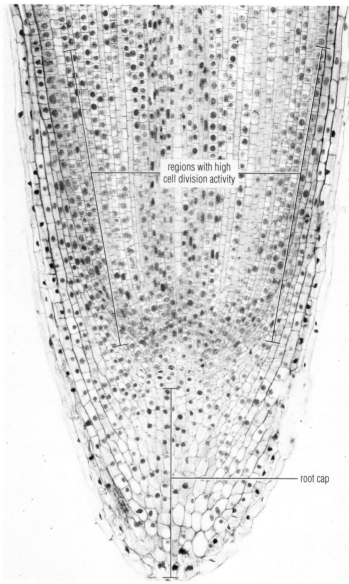

regions with high cell division activity

root cap

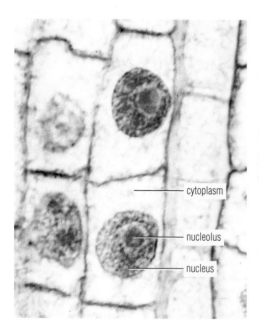

cytoplasm

nucleolus

nucleus

Figure 18b **Interphase** cells prior to mitosis (l.s., 1400×). (Photo by J. W. Perry)

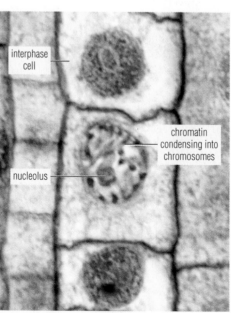

interphase cell

chromatin condensing into chromosomes

nucleolus

Figure 18c **Early prophase**. Nuclear envelope intact (l.s., 1400×). (Photo by J. W. Perry)

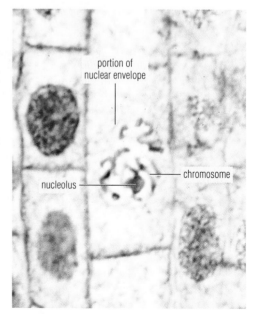

portion of nuclear envelope

chromosome

nucleolus

Figure 18d **Later prophase**. Nuclear envelope disorganizing (l.s., 1400×). (Photo by J. W. Perry)

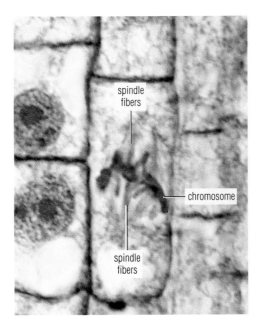

Figure 19a Metaphase. Duplicated chromosomes (each consisting of two chromatids) at equatorial plane (l.s., 1400×). (Photo by J. W. Perry)

Figure 19b Anaphase. Sister chromatids separating into unduplicated (daughter) chromosomes and moving toward opposite poles (l.s., 1400×). (Photo by J. W. Perry)

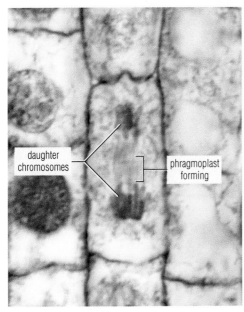

Figure 19c Telophase. Unduplicated (daughter) chromosomes at poles (l.s., 1400×). (Photo by J. W. Perry)

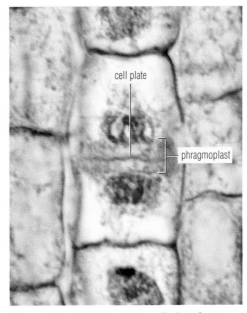

Figure 19d Cytokinesis by **cell plate formation** (l.s., 1400×). (Photo by J. W. Perry)

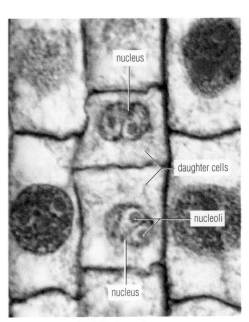

Figure 19e Two **daughter cells** following cytokinesis. Note that both the nuclei and cytoplasmic volume in these two newly formed cells are smaller than older neighboring cells (l.s., 1400×). (Photo by J. W. Perry)

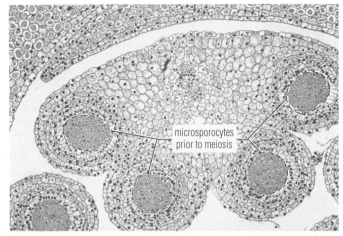

Figure 20a **Lily** (*Lilium*) **anther**. Nuclei of certain cells (microsporocytes) within the anther undergo meiosis to become pollen grains (immature male plants that produce haploid [*n*] sperm). Figures 20b–21g are different stages of meiosis taken from within this region (prep. slide, c.s., 52×). (Photo by J. W. Perry)

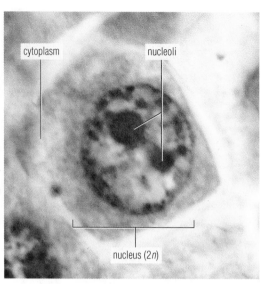

Figure 20b Diploid (2*n*) **interphase** cells (sec., 1800×). (Photo by J. W. Perry)

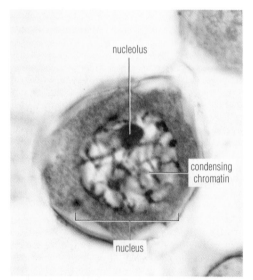

Figure 20c **Early prophase I** (sec., 1200×). (Photo by J. W. Perry)

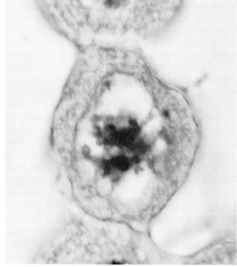

Figure 20d **Mid-prophase I** (sec., 1200×). (Photo by J. W. Perry)

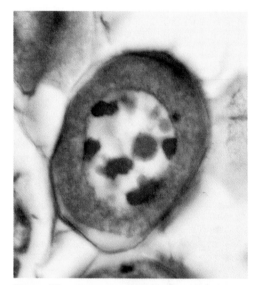

Figure 20e **Late prophase I** (sec., 1200×). (Photo by J. W. Perry)

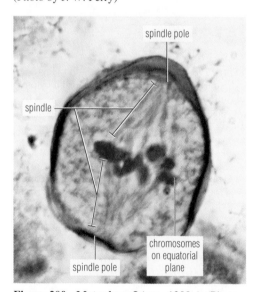

Figure 20f **Metaphase I** (sec., 1200×). (Photo by J. W. Perry)

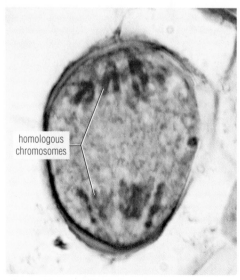

Figure 20g **Anaphase I**. The homologous chromosomes are separated (sec., 1200×). (Photo by J. W. Perry)

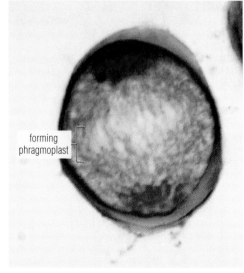

Figure 20h **Telophase I** (sec., 1200×). (Photo by J. W. Perry)

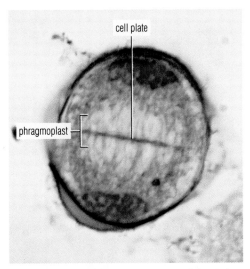

Figure 21a   Meiosis I complete; **cytokinesis** by **cell plate formation** beginning (sec., 1200×). (Photo by J. W. Perry)

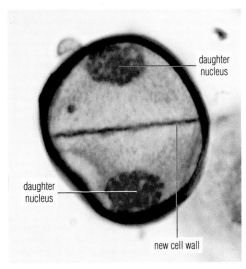

Figure 21b   Cytokinesis complete (sec., 1200×). (Photo by J. W. Perry)

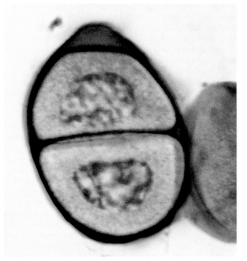

Figure 21c   Prophase II (sec., 1200×). (Photo by J. W. Perry)

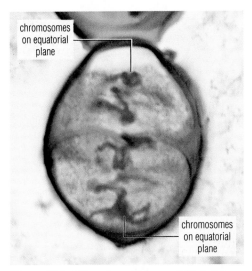

Figure 21d   Metaphase II (sec., 1200×). (Photo by J. W. Perry)

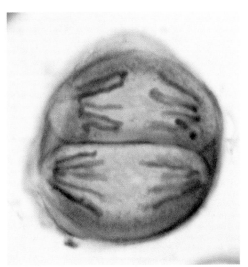

Figure 21e   Anaphase II. Sister chromatids have separated into unduplicated daughter chromosomes (sec., 1200×). (Photo by J. W. Perry)

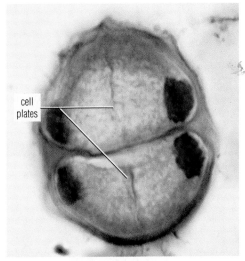

Figure 21f   Telophase II and **cytokinesis** beginning in both cells (sec., 1200×). (Photo by J. W. Perry)

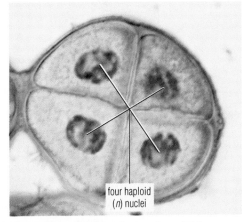

Figure 21g   Cytokinesis complete. All four cells are haploid (n). Following this stage, the four cells separate and differentiate, becoming pollen grains. Each pollen grain eventually produces haploid sperm (sec., 1200×). (Photo by J. W. Perry)

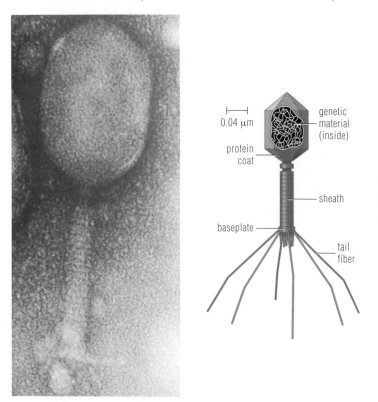

Figure 22a   Transmission electron micrograph (left) and artist interpretation (right) of a **virus** that infects a bacterial cell (called a "**bacteriophage**") (t.e.m., 530,000×). (Photo courtesy the Perkin-Elmer Corporation)

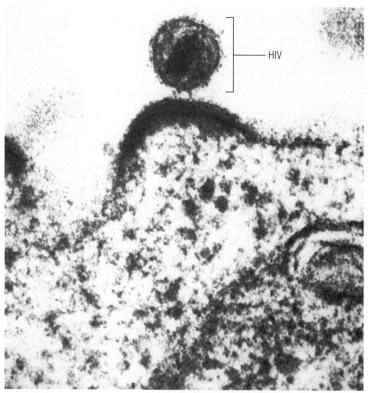

Figure 22b   Transmission electron micrograph of **HIV** (*H*uman *I*mmunodeficiency *V*irus), the infective agent triggering **AIDS** (*A*cquired *I*mmune *D*eficiency *S*yndrome) (t.e.m., 500,000×). (Photo courtesy Department of Health and Human Services, National Institute of Allergy and Infectious Diseases)

Figure 22c   Gladiolus break, a disease caused by a virus that causes streaking of the flowers and mottling of the leaves (live, 0.5×). (Photo by J. W. Perry)

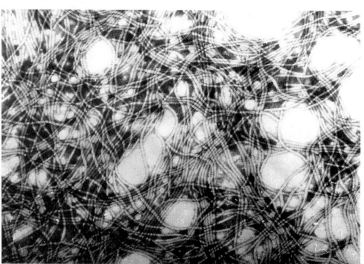

Figure 22d   Transmission electron micrograph of a virus like that which causes gladiolus break. **Virions** (virus particles) are the numerous strands (t.e.m., 140,000×). (Photo courtesy University of Wisconsin–Madison, Department of Plant Pathology)

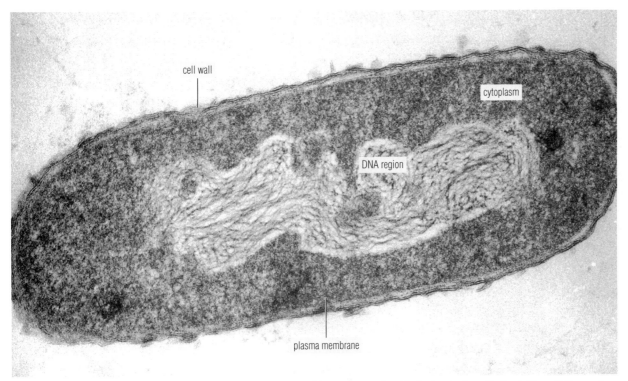

cell wall

cytoplasm

DNA region

plasma membrane

Figure 23a   Transmission electron micrograph of a **heterotrophic prokaryotic cell**, the **bacterium** *Escherichia coli* (t.e.m., l.s., 63,500×). (Photo courtesy G. Cohen-Bazire)

Figure 23b   Bacilli (rod-shaped) **bacteria** from a sewage treatment sample (live, w.m., d.i.c. microscopy, 1350×). (Photo by J. W. Perry)

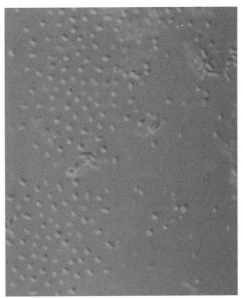

Figure 23c   Cocci (spherical) **bacteria** from a sewage treatment sample (live, w.m., d.i.c. microscopy, 1350×). (Photo by J. W. Perry)

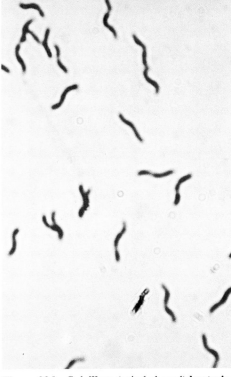

Figure 23d   Spirillum (spiral-shaped) **bacteria** *Spirillum volutans*, a particularly large species (live, w.m., 1900×). (Photo by J. W. Perry)

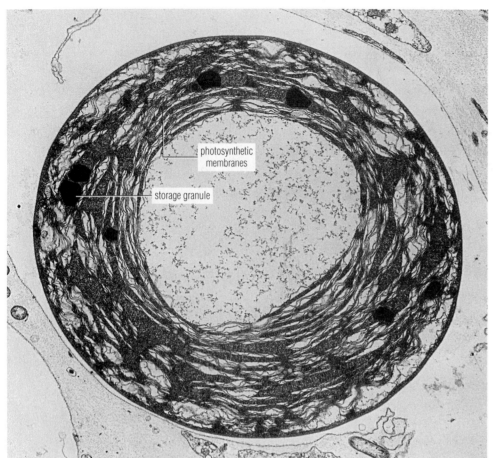

Figure 24a Transmission electron micrograph of *Prochloron*, a photosynthetic bacterium that has the same pigments found in green algae and plants (t.e.m., sec., 11,000×). (Photo by T. J. Pugh and E. H. Newcomb, University of Wisconsin–Madison)

photosynthetic membranes

storage granule

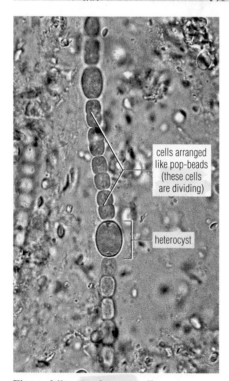

Figure 24b *Anabaena azollae*, a **cyanobacterium** that lives as a symbiont within the leaves of the water fern *Azolla* (Figure 69b). Enlarged cells are **hetero-cysts**, functioning in nitrogen fixation (live, w.m. from a crushed *Azolla* leaf, 600×). (Photo by J. W. Perry)

cells arranged like pop-beads (these cells are dividing)

heterocyst

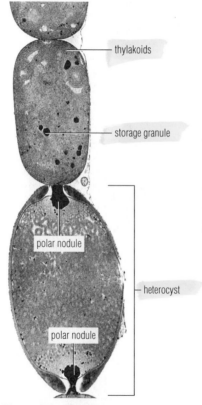

Figure 24c Transmission electron micrograph of *Anabaena* cells (t.e.m., l.s., 13,000×). (Photo by R. D. Warmbrodt)

thylakoids

storage granule

polar nodule

heterocyst

polar nodule

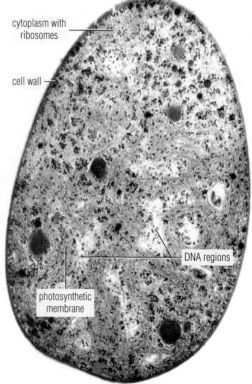

Figure 24d Transmission electron micrograph of a single **photosynthetic prokaryotic cell**, the **cyanobacterium** *Anabaena* (t.e.m., l.s., 10,500×). (Photo by R. D. Warmbrodt)

cytoplasm with ribosomes

cell wall

DNA regions

photosynthetic membrane

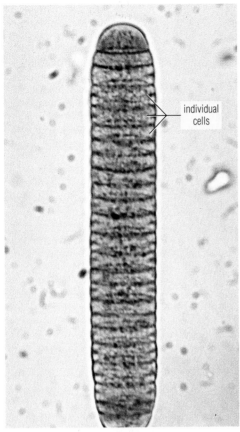

Figure 25a A short filament of *Oscillatoria* (live, w.m., 1200×). (Photo by J. W. Perry)

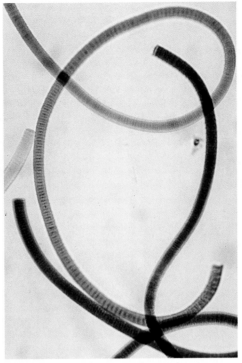

Figure 25b *Oscillatoria* filaments (prep. slide, w.m., 350×). (Photo by J. W. Perry)

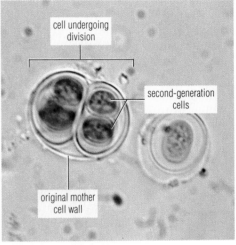

Figure 25c *Gleocapsa* (live, w.m., 1300×). (Photo by J. W. Perry)

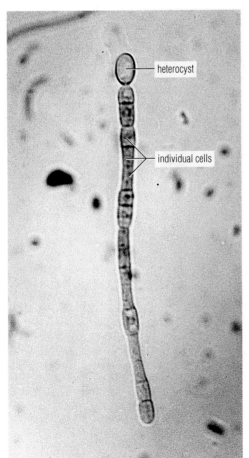

Figure 25d *Gloeotrichia*, with a terminal heterocyst (live, w.m., 630×). (Photo by J. W. Perry)

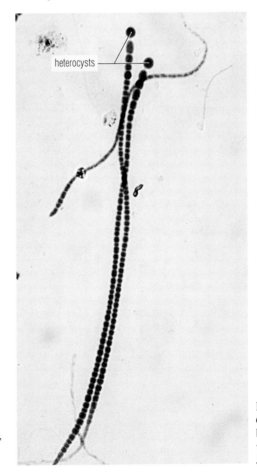

Figure 25e *Gloeotrichia* (**cyanobacterium**) (prep. slide, w.m., 550×). (Photo by J. W. Perry)

Figure 26a *Allomyces arbuscula* **mitosporangium** containing diploid (2*n*) **zoospores** that have been produced by mitosis (live, w.m. 700×). (Photo by J. W. Perry)

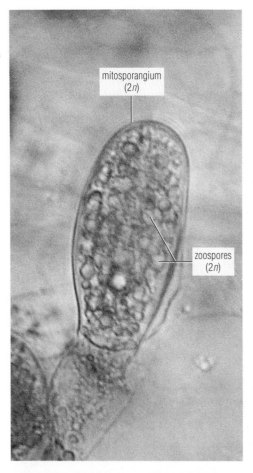

mitosporangium (2*n*)

zoospores (2*n*)

Figure 26b *Allomyces arbuscula* **meiosporangium** (resting sporangium) containing haploid (*n*) **zoospores** that have been produced by meiosis. Note the prominent pitting in the sporangial wall (live, w.m., 700×). (Photo by J. W. Perry)

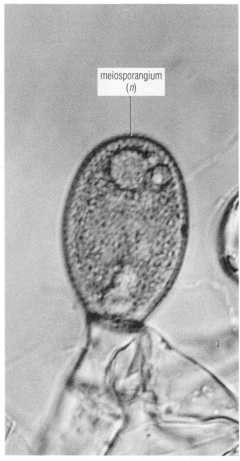

meiosporangium (*n*)

Figure 26c *Allomyces arbuscula* **gametangia** containing haploid (*n*) **gametes** (live, w.m., 600×). (Photo by J. W. Perry)

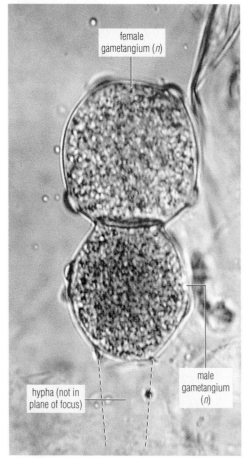

female gametangium (*n*)

hypha (not in plane of focus)

male gametangium (*n*)

diseased (infected) tubers

healthy (uninfected) tubers

Figure 26d Black wart of potato disease, caused by the chytrid *Synchytrium endobioticum* (live, 0.2×). (Photo by E. Wade)

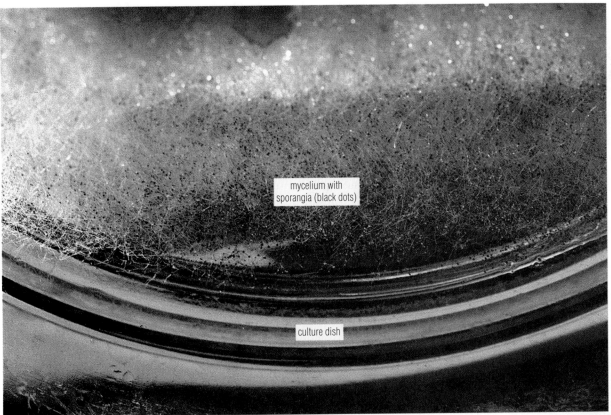

mycelium with sporangia (black dots)

culture dish

Figure 27a Mycelium of the **bread mold *Rhizopus*** growing on a slice of bread. The black dots are **sporangia** on the mycelium (live, 1×). (Photo by J. W. Perry)

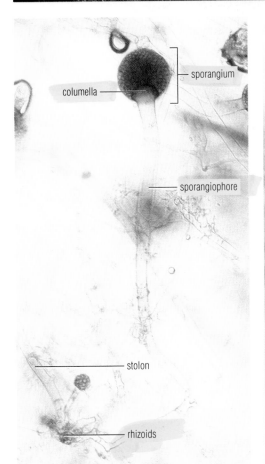

sporangium

columella

sporangiophore

stolon

rhizoids

Figure 27b **Sporangium** of ***Rhizopus*** atop sporangiophore (live, w.m., 130×). (Photo by J. W. Perry)

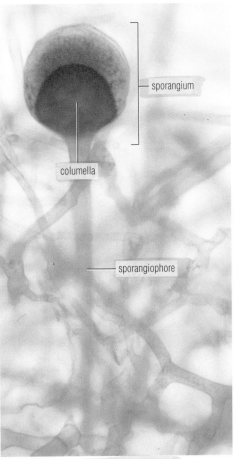

sporangium

columella

sporangiophore

Figure 27c ***Rhizopus* sporangium**. In this stained preparation, the columella is apparent (prep. slide, w.m., 165×). (Photo by J. W. Perry)

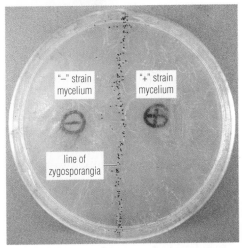

"–" strain mycelium

"+" strain mycelium

line of zygosporangia

Figure 27d Culture plate of a **bread mold** (***Phycomyces***) undergoing sexual reproduction. The vertical black line consists of **zygosporangia** formed as a result of fusion of gametangia produced by the "+" strain with gametangia produced by the "–" strain (live, 0.5×). (Photo by J. W. Perry)

Know

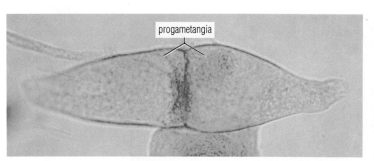

Figure 28a   Sexual reproduction in *Rhizopus*. **Progametangia** of two strains that have met (prep. slide, w.m., 300×). (Photo by J. W. Perry)

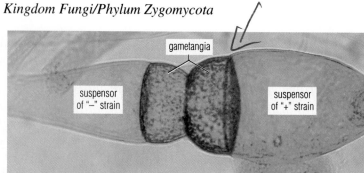

Figure 28b   Sexual reproduction in *Rhizopus*. **Gametangia** have been produced from progametangia as a result of concentration of cytoplasm and subsequent formation of cell walls (prep. slide, w.m., 300×). (Photo by J. W. Perry)

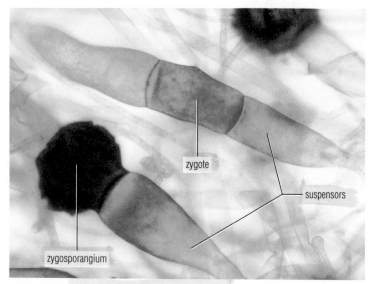

Figure 28c   Sexual reproduction in *Rhizopus*. **Zygote** formed by fusion of nuclei from respective gametangia after the walls at their tips broke down and the cytoplasms fused. Eventually, a thick wall forms around the zygote, producing the **zygosporangium** (prep. slide, w.m., 300×). (Photo by J. W. Perry)

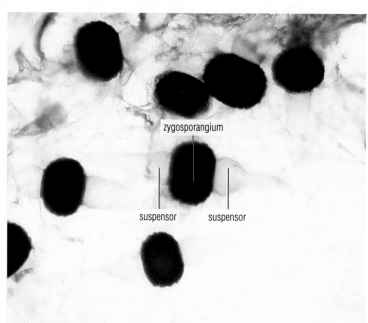

Figure 28d   Living zygosporangia of *Rhizopus* (live, w.m., 120×). (Photo by J. W. Perry)

Figure 28e   The gun fungus *Pilobolus*, growing on horse dung. Sporangia at the tips of the enlarged bulbs (subsporangial vesicles) seen here may be shot up to 6 meters by a hydraulic mechanism (live, 1.1×). (Photo by J. W. Perry)

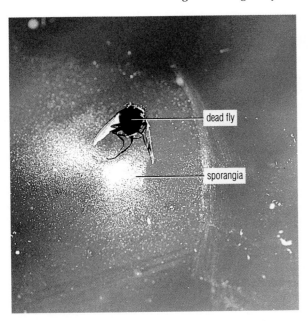

Figure 29a Halo of **sporangia** of the **fly fungus** *Entomophthora museae*. This fungus infects and kills flies and is often found stuck on window glass (live, 1.5×). (Photo by J. W. Perry)

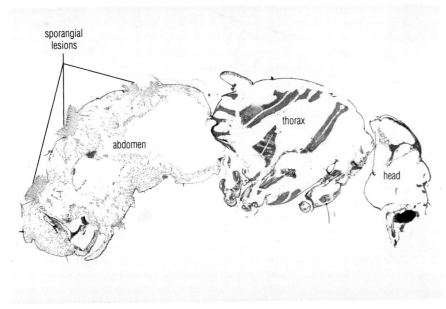

Figure 29b Longitudinal section of fly that has been infected by the **fly fungus** *Entomophthora museae* (prep. slide, l.s., 14×). (Photo by J. W. Perry)

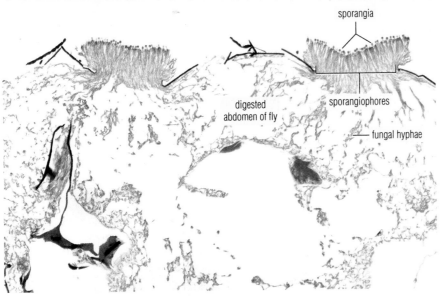

Figure 29c Sporangia and sporangiophores of the **fly fungus** *Entomophthora museae* emanating from the fly body. These sporangia are picked up by and infect living flies (prep. slide, l.s., 50×). (Photo by J. W. Perry)

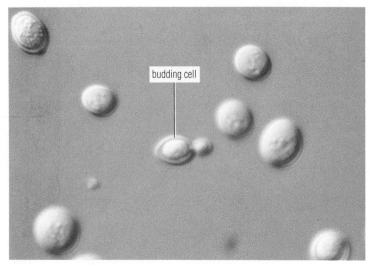

Figure 30a   Cells of the **baker's yeast** *Saccharomyces*. One of these cells is reproducing asexually by **budding** (live, w.m., d.i.c. microscopy, 1700×). (Photo by J. W. Perry)

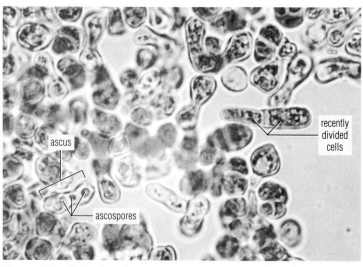

Figure 30b   Cells of the **fission yeast** *Schizosaccharomyces octosporous*. The yeast reproduces asexually by fission, and sexually by **ascospores** that are contained in an **ascus** (live, w.m., 1044×). (Photo by J. W. Perry)

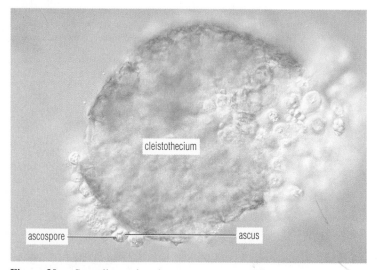

Figure 30c   Lilac (*Syringa vulgaris*) leaf infected with the **powdery mildew** *Microsphaera*. The white mycelium covers the leaf and produces chains of **conidia** (live, 1×). (Photo by J. W. Perry)

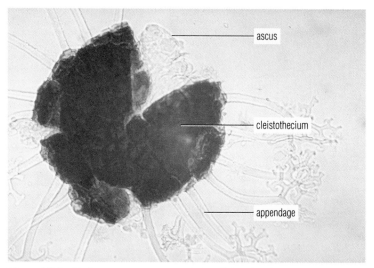

Figure 30d   Broken-open ascus-containing **cleistothecium** of the **powdery mildew** *Microsphaera*. This appendaged sphere is the product of sexual reproduction and can be found as many "black dots" on the leaves of lilacs in late summer (live, w.m., 90×). (Photo by J. W. Perry)

Figure 30e   Sexually produced, ascus-containing **cleistothecium** (a type of ascocarp) of the **blue mold** *Eurotium chevalieri* that has been crushed to release the asci (live, w.m., d.i.c. microscopy, 400×). (Photo by J. W. Perry)

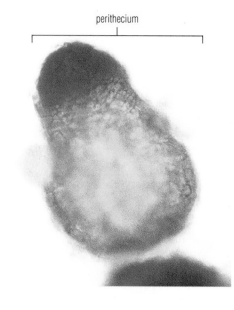

Figure 30f
Sexually produced, ascus-containing **perithecium** (a type of ascocarp) of the **sac fungus** *Sordaria* (live, w.m., 175×). (Photo by J. W. Perry)

Figure 31a   Sexually produced, ascus-containing **apothecium** (a third type of ascocarp) of the **cup fungus** *Peziza*. The asci line the interior of the cup (live, 0.3×). (Photo by J. W. Perry)

Figure 31b
Sexually produced, ascus-containing **apothecia** (ascocarps) of the highly prized *Morchella*, a spring "mushroom" commonly called a **morel**. The asci line the ridges of this highly modified apothecium. These fungi are a true culinary delight (live, 0.4×). (Photo by J. W. Perry)

Figure 31c
Apothecia of two species of *Morchella* (*M. esculenta* on the left and *M. semilibra* on the right) that have been cut longitudinally to show their hollow nature and attachment of the cap to the stalk. These features are used to distinguish morels from the less desirable and sometimes poisonous false morels shown in Figure 31f (live, l.s., 0.5×). (Photo by J. W. Perry)

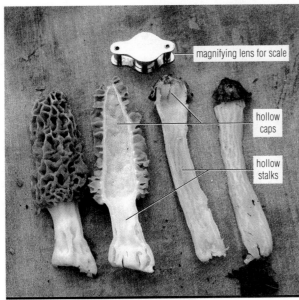
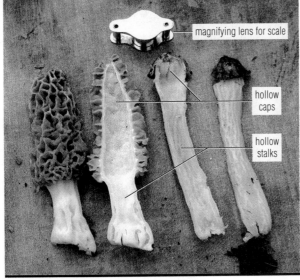

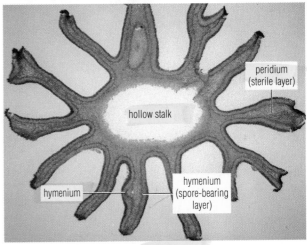

Figure 31d   Cross section of *Morchella* **apothecium** (ascocarp) with its two identifiable layers: the ascospore-containing hymenium and the sterile peridium (prep. slide, c.s., 20×). (Photo by J. W. Perry)

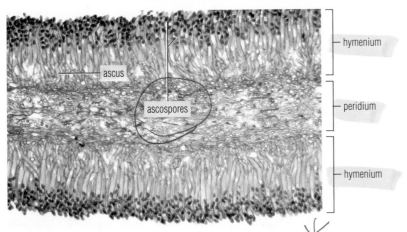

Figure 31e   **Asci** (singular, ascus) containing **ascospores** of *Morchella* (prep. slide, c.s. 250×). (Photo by J. W. Perry)

Figure 31f
Apothecium (ascocarp) of the **false morel**, *Gyromitra*. Some of these species are poisonous. They are distinguished from the true morels by the form of attachment of the cap to the stalk. Compare with Figures 31b and 31c (live, 0.25×). (Photo by J. W. Perry)

Figure 32a **Apothecium** (ascocarp) of a **truffle** (*Tuber*), the most expensive and highly prized of the edible fungi. These subterranean ascocarps are hunted with dogs and/or pigs in France and Italy and fetch about $600 per pound on the consumer market (live, 1×). (Photo by J.-C. Davidian and G. Callot)

Figure 32b **Apothecium** (ascocarp) of a **truffle** (*Tuber*) showing the **ascospores** (prep. slide, sec., 220×). (Photo by J. W. Perry)

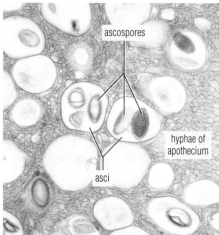

Figure 32c **Apple scab disease** caused by the ascomycete *Venturia inaequalis*. The fungus infects both the fruit (shown here) and the leaves, as shown in Figures 32d and 32e (live, 1×) (Photo by J. W. Perry)

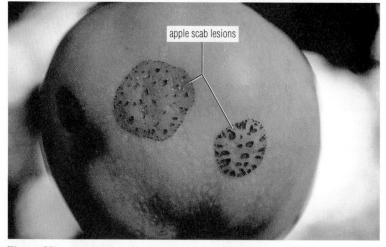

Figure 32d Apple leaf infected by *Venturia inaequalis* that is reproducing asexually by **conidia** (prep. slide, c.s., 200×). (Photo by Triarch Microslides, Inc.)

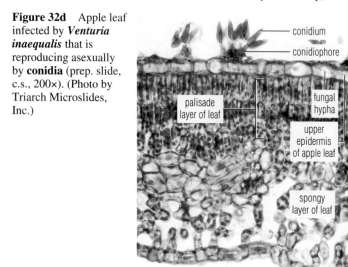

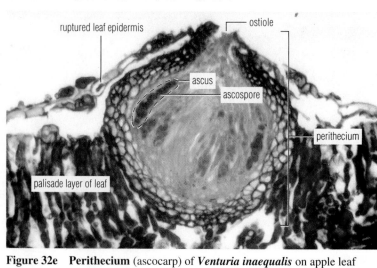

Figure 32e **Perithecium** (ascocarp) of *Venturia inaequalis* on apple leaf (prep. slide, l.s., 400×). (Photo courtesy Triarch Microslides, Inc.)

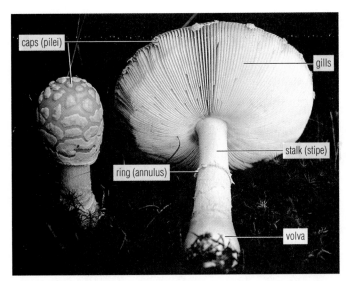

Figure 33a   Sexually produced **basidiocarps** of the **club fungus** *Amanita muscaria*. The one on the left is younger, its cap not yet fully expanded (live, 0.5×). (Photo by J. W. Perry)

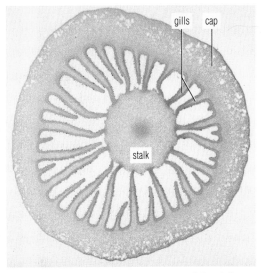

Figure 33b   Cross section of the **cap** of the **club fungus** *Coprinus* (prep. slide, c.s., 16×). (Photo by J. W. Perry)

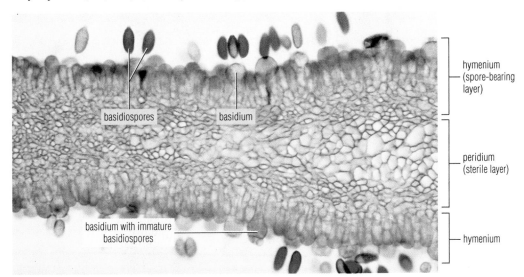

Figure 33c   *Coprinus* **gill** with **basidia-bearing basidiospores**. In this fungus, each basidium produces four basidiospores. The commercial mushroom commonly sold in grocery stores is *Agaricus bisporus*; as the specific epithet suggests, it produces only two spores per basidium (prep. slide, c.s., 520×). (Photo by J. W. Perry)

Figure 33d   A **bolete** (*Leccinum scabrum*). Boletes are fleshy club fungi, but their basidiospores are produced in pores (live, 0.5×) (Photo by J. W. Perry)

Figure 33e   **Shelf fungus** *Ganoderma*. This woody basidiocarp, called a **conk**, is an indication of mycelium growing within the tree (live, 0.1×). (Photo by J. W. Perry)

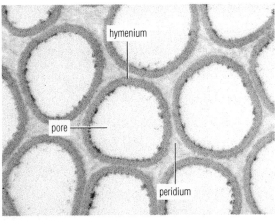

Figure 33f   Cross section of lower surface of a **conk**. The basidia line the edges of the pores. Red "dots" are basidiospores (prep. slide, c.s., 100×). (Photo by J. W. Perry)

Figure 34a   Basidiocarp of a **coral fungus** (*Clavulina*) (live, 0.75×). (Photo by J. W. Perry)

Figure 34b   Basidiocarp of a **tooth fungus** (*Hericium*) (live, 0.75×). (Photo by J. W. Perry)

Figure 34c   Mature basidiocarps of **puffballs**. The pore at the top of the puffballs allows the internally produced basidiospores to escape (live, 0.5×). (Photo by J. W. Perry)

Figure 34d   Immature basidiocarps of **puffballs** (*Calvatia*). Before the basidiospores develop, the flesh is firm, as illustrated in the one that has been sliced open (live, 0.5×). (Photo by J. W. Perry)

Figure 34e   Basidiocarp of an **earth star** (*Geastrum*) (live, 0.5×). (Photo by J. W. Perry)

Figure 35a Basidiocarp of the **stinkhorn fungus** (*Phallus*). The flesh of this fungus gives off an odor disagreeable to most humans but attractive to flies, which pick up the basidiospores and carry them to new environments (live, 0.5×). (Photo by J. W. Perry)

gleba

fly picking up basidiospores

stalk

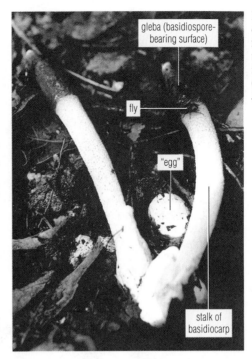

gleba (basidiospore-bearing surface)

fly

"egg"

stalk of basidiocarp

Figure 35b Basidiocarp of the **stinkhorn fungus** (*Phallus*). The unelongated basidiocarp is referred to as an "egg" (live, 0.5×). (Photo by J. W. Perry)

Figure 35c Basidiocarps of a **bird's-nest fungus** (*Cyathus*). The "eggs" contain the basidiospores (live, 1×). (Photo by J. W. Perry)

rotting wood

"eggs" (spore-bearing structures)

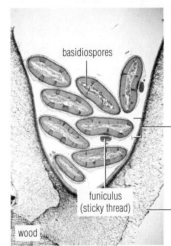

basidiospores

funiculus (sticky thread)

peridiole

basidiocarp

wood

Figure 35d Basidiocarp of a **bird's-nest fungus** (*Cyathus*) showing basidiospore-containing peridioles ("eggs") (prep. slide, l.s., 9×). (Photo by J. W. Perry)

basidiocarp

Figure 35e Basidiocarp of a **jelly fungus** (*Exidia*), so called because of its gelatinous consistency (live, 0.5×). (Photo by J. W. Perry)

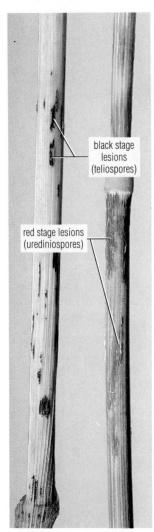

Figure 36a   Symptoms of **wheat rust disease** caused by *Puccinia graminis* (live., 1×). (Photo by C. R. Grau and S. Vicen)

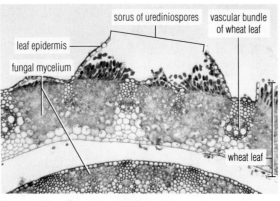

Figure 36b   *Puccinia graminis*, the organism causing **wheat rust disease**. **Urediniospores** (uredospores or urediospores) have erupted from the surface of a wheat leaf. This is the **red** (uredial) **stage** of the disease, produced during North American summers (prep. slide, c.s., 20×). (Photo by J. W. Perry)

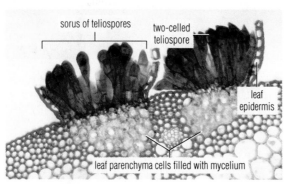

Figure 36d   *Puccinia graminis* **teliospores** erupting from the surface of a wheat leaf. This is the **black** (telial), **overwintering stage** of the disease (prep. slide, c.s., 230×). (Photo by J. W. Perry)

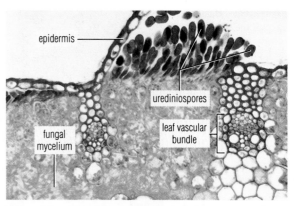

Figure 36c   *Puccinia graminis* **urediniospores** growing from a mycelium-filled wheat leaf (prep. slide, c.s., 230×). (Photo by J. W. Perry)

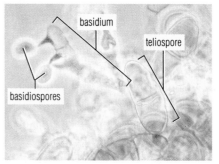

Figure 36e   Germinating **teliospores** of another rust fungus (*Gymnosporangium*). The teliospores have produced basidia on which basidiospores are borne (live, w.m., 350×). (Photo by J. W. Perry)

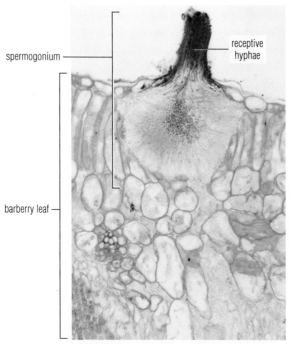

Figure 36f   *Puccinia graminis* **spermogonium** (pycnium) on a barberry leaf, the alternate host of the pathogen (prep. slide, c.s., 260×). (Photo by J. W. Perry)

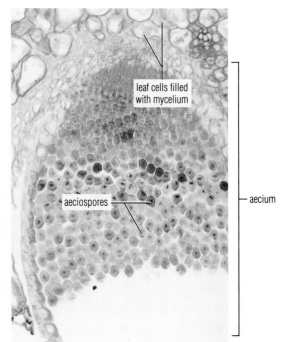

Figure 36g   *Puccinia graminis* **aecium** containing **aeciospores** on a barberry leaf (prep. slide, c.s., 260×). (Photo by J. W. Perry)

Figure 37b   Telia gall (a woody profusion of juniper stem tissue caused by the infection) with gelatinous "horns" bearing basidia and basidiospores (live, 0.5×). (Photo by J. W. Perry)

Figure 37a   Cedar-apple rust disease caused by *Gymnosporangium* on a juniper (*Juniperus*) tree. This is the **telial stage**, which becomes obvious after prolonged warm rain that stimulates the production of the orange "telial horns" (live, 0.01×). (Photo by J. W. Perry)

Figure 37d   Spermagonial (pycnial) and **aecial stages** of **cedar-apple rust** on the leaves of hawthorn (*Crataegus*) (live, 0.5×). (Photo by J. W. Perry)

Figure 37c   Cedar-apple rust on its alternate host, hawthorn (*Crataegus*). This tree died the year after the photo was taken (live, 0.05×). (Photo by J. W. Perry)

Figure 37e   Corn smut disease caused by *Ustilago maydis*. The gray basidiospore-containing structures are corn kernels that have enlarged as a result of the infection. They may appear disgusting but are edible (live, 0.5×). (Photo by J. W. Perry)

Figure 38a A variety of **molds** colonizing squash. These white **mycelia** covering the squash are shedding literally billions of blue-colored asexual **conidia**, their only means of reproduction (live, 0.15×). (Photo by J. W. Perry)

Figure 38b *Aspergillus* **conidiophores** bearing **conidia** (prep. slide, w.m., 200×). (Photo by Biodisc, Inc.)

Figure 38c *Aspergillus* **conidiophore** and **conidia** (live, w.m., 375×). (Photo by J. W. Perry)

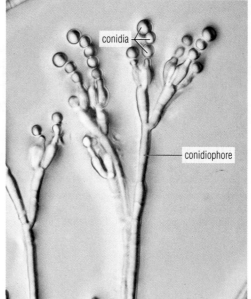

Figure 38d *Penicillium* **conidiophore** and **conidia** (live, w.m., d.i.c. microscopy, 925×). (Photo by J. W. Perry)

Figure 38e *Alternaria* **conidia**. These airborne multicellular conidia cause an allergic reaction in many humans (live, w.m., 370×). (Photo by J. W. Perry)

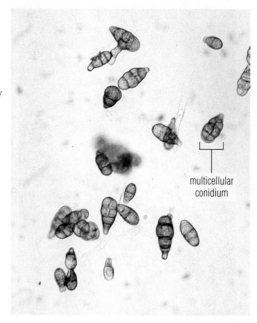

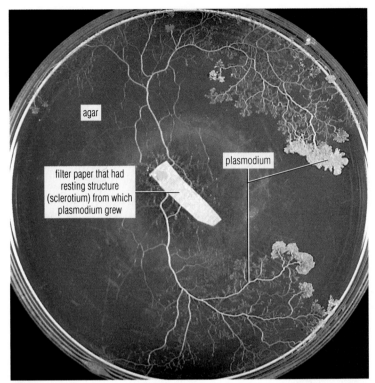

Figure 39a   Culture dish containing the **plasmodial slime mold** *Physarum polycephalum* (1×). (Photo by J. W. Perry)

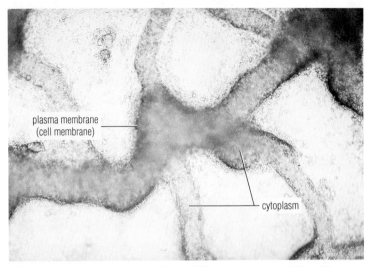

Figure 39b   Portion of **plasmodium** of *Physarum polycephalum* (live, 90×). (Photo by J. W. Perry)

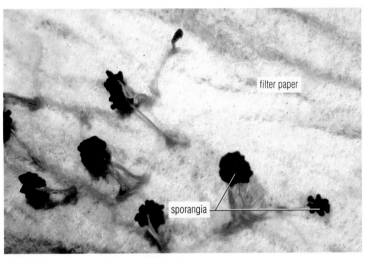

Figure 39c   Spore-containing **sporangia** of *Physarum polycephalum* (live, 1.5×). (Photo by J. W. Perry)

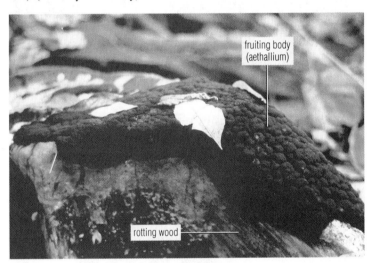

Figure 39d   Spore-containing **fruiting body** (an aethallium) of the **slime mold** *Badhemia maxima*. Slime molds like this one are common on decaying wood and bark mulch, often appearing brightly colored before producing spores (live, 0.5×). (Photo by J. W. Perry)

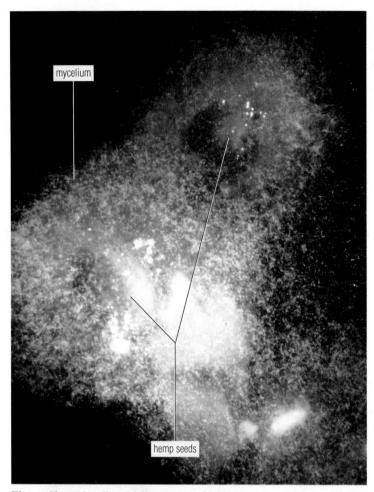

Figure 40a Mycelium of the **water mold** *Saprolegnia* growing on hemp seeds. The numerous white dots are oogonia (live, 4.5×). (Photo by J. W. Perry)

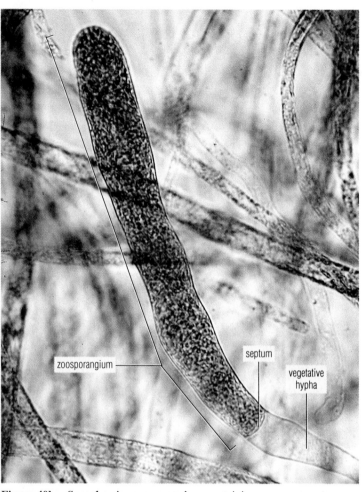

Figure 40b *Saprolegnia* **zoosporangium** containing numerous zoospores (live, w.m., 450×). (Photo by C. A. Taylor III)

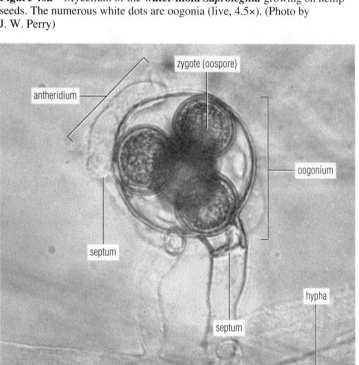

Figure 40c *Saprolegnia* **gametangia**, the **oogonium** (female gametangium) containing **zygotes** (oospores) produced when sperm from the **antheridium** (male gametangium) fertilized the egg cells (oospheres) (live, w.m., 550×). (Photo by J. W. Perry)

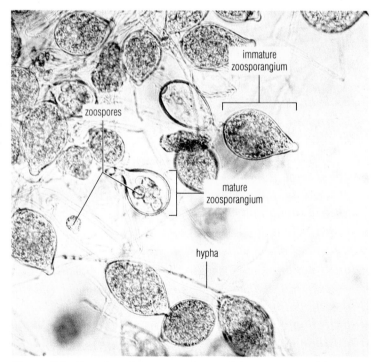

Figure 40d Pear-shaped **zoosporangia** containing **zoospores** of the **water mold** *Phytophthora cactorum* (live, w.m., 340×). (Photo by J. W. Perry)

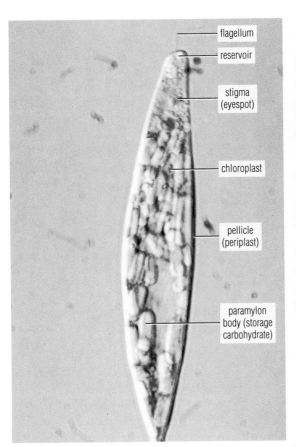

Figure 41a labels: flagellum, reservoir, stigma (eyespot), chloroplast, pellicle (periplast), paramylon body (storage carbohydrate)

Figure 41a *Euglena ascus* (live, w.m., d.i.c. microscopy, 1100×). (Photo by J. W. Perry)

Figure 41b Scanning electron micrograph of *Euglena*. The scanning electron microscope is used to resolve surface features of organisms. Here the **pellicle (periplast)** and **flagellum** are obvious (s.e.m., w.m., 825×). (Photo from Shih/Kessel, *Living Images*, ©1982 Science Books International; reprinted with permission of present publisher, Jones and Bartlett Publishers)

Figure 41b labels: flagellum, pellicle (periplast)

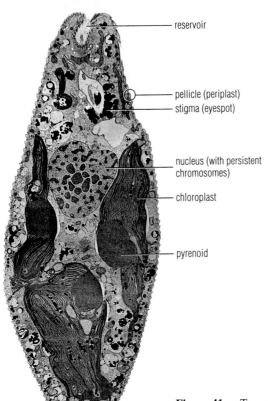

Figure 41c labels: reservoir, pellicle (periplast), stigma (eyespot), nucleus (with persistent chromosomes), chloroplast, pyrenoid

Figure 41c Transmission electron micrograph of *Euglena* (t.e.m., l.s., 1240×). (Photo by P. L. Walne and H. J. Arnott)

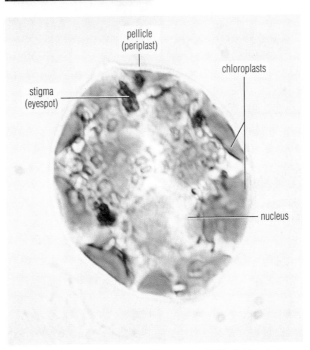

Figure 41d labels: pellicle (periplast), chloroplasts, stigma (eyespot), nucleus

Figure 41d *Trachelomonas* (live, w.m., 1500×). (Photo by J. W. Perry)

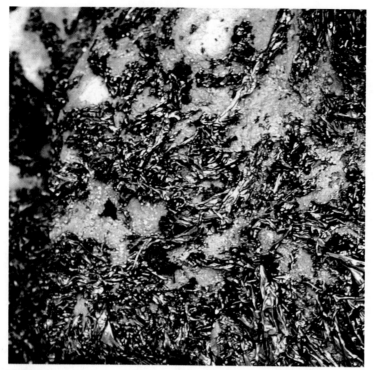

Figure 42a   Marine **red alga**, *Porphyra*, at low tide. This organism is widespread throughout the world's oceans and is sold dried in food stores as **nori** (live, 0.5×). (Photo by J. W. Perry)

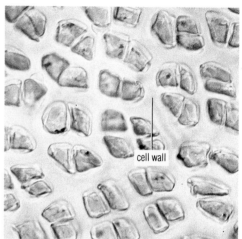

Figure 42b   Microscopic appearance of *Porphyra* (w.m., dried specimen, 370×). (Photo by J. W. Perry)

Figure 42c   Pacific Coast **red alga** (*Iridaea*), with **coralline algae** (*Corallinia*) attached (live, 0.3×). (Photo by J. W. Perry)

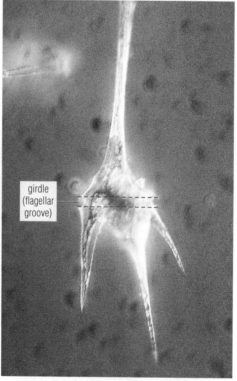

Figure 42d   The **dinoflagellate** *Ceratium*. The dashed lines define the limits of the flagellar groove (prep. slide, w.m., d.i.c. microscopy, 410×). (Photo by J. W. Perry)

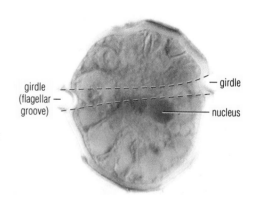

Figure 42e   The **dinoflagellate** *Peridinium*. The dashed lines define the limits of the flagellar groove (prep. slide, w.m., d.i.c. microscopy, 1000×). (Photo by J. W. Perry)

Figure 43a **Diatomaceous earth** quarry near Quincy, Washington. (Photo courtesy Dan Williams)

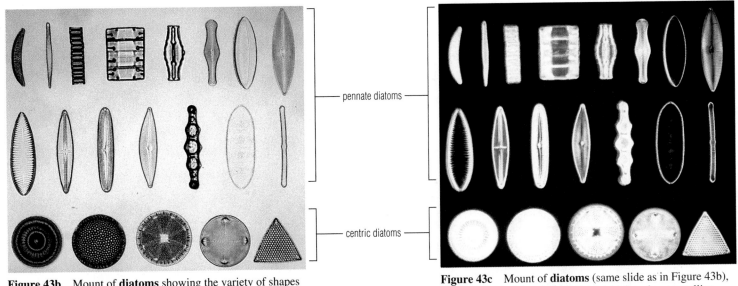

pennate diatoms

centric diatoms

Figure 43b Mount of **diatoms** showing the variety of shapes of different species (prep. slide, w.m., 95×). (Photo by J. W. Perry)

Figure 43c Mount of **diatoms** (same slide as in Figure 43b), photographed with polarized light, which makes crystalline structures appear to "glow" against a dark background (95×). (Photo by J. W. Perry)

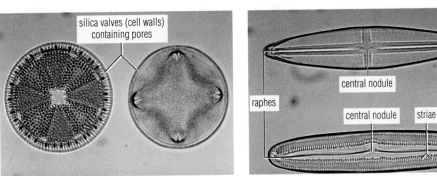

silica valves (cell walls) containing pores

Figure 43d Valve view of two **centric** (radially symmetrical) **diatoms**, typical of those found in marine (saltwater) environments (prep. slide, w.m., 190×). (Photo by J. W. Perry)

central nodule

raphes

central nodule   striae   polar nodule

Figure 43e Valve view of two **pennate** (bilaterally symmetrical) **diatoms**, typical of those found in freshwater environments (prep. slide, w.m., 235×). (Photo by J. W. Perry)

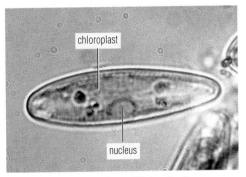

chloroplast

nucleus

Figure 43f Valve view of living freshwater **diatom** (live, w.m., 235×). (Photo by J. W. Perry)

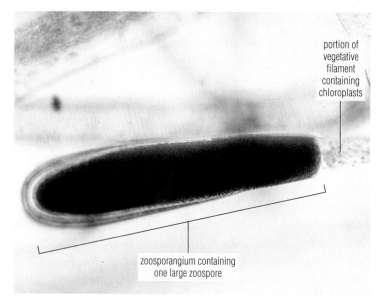

portion of
vegetative
filament
containing
chloroplasts

zoosporangium containing
one large zoospore

Figure 44a   *Vaucheria* **zoosporangium** (live, w.m., 270×). (Photo by J. W. Perry)

Figure 44b   Single large **zoospore** of *Vaucheria* shortly after being shed from zoosporangium (live, w.m., 380×). (Photo by J. W. Perry)

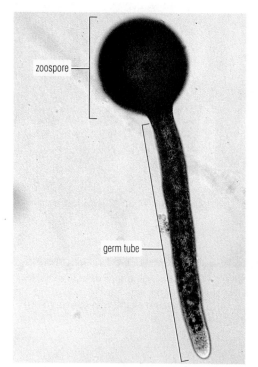

zoospore

germ tube

Figure 44c   Germinating **zoospore** of *Vaucheria* (live, w.m., 160×). (Photo by J. W. Perry)

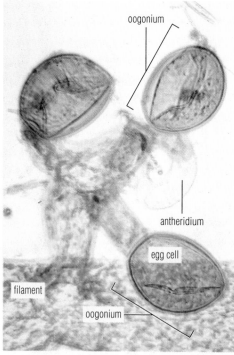

oogonium

antheridium

egg cell

filament

oogonium

Figure 44d   **Sex organs** (gametangia) of *Vaucheria geminata* (prep. slide, w.m., 370×). (Photo by J. W. Perry)

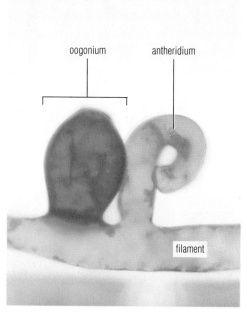

oogonium

antheridium

filament

Figure 44e   **Sex organs** (gametangia) of *Vaucheria sessilis* (prep. slide, w.m., 290×). (Photo by J. W. Perry)

Figure 45a *Laminaria*, a marine **brown alga** found along the northern Atlantic and Pacific coasts of the United States. Large marine brown algae are called **kelps** (live, 0.06×). (Photo by J. W. Perry)

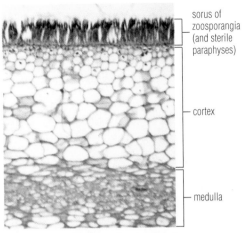

Figure 45b Portion of the **blade** of *Laminaria* showing reproductive structures (prep. slide, sec., 65×). (Photo by J. W. Perry)

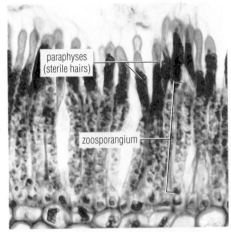

Figure 45c **Zoosporangia** of *Laminaria* (prep. slide, c.s., 260×). (Photo by J. W. Perry)

Figure 45d Colony of the Pacific **sea palm**, *Postelsia* (live, 0.04×). (Photo by J. W. Perry)

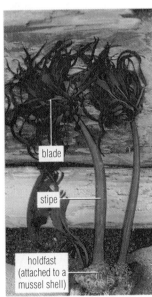

Figure 45e *Postelsia*, the **sea palm** (live, 0.4×). (Photo by J. W. Perry)

Figure 46a *Macrocystis*, a **kelp** found along the Pacific Coast of the United States. This kelp has been reported to grow to a length of up to 500 meters (about 1500 feet) (live, 0.04×). (Photo by J. W. Perry)

floats

stipe

blade

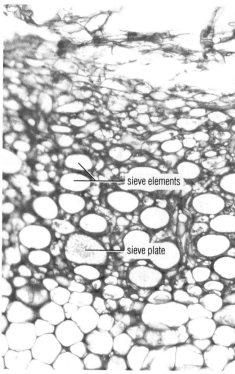

sieve elements

sieve plate

Figure 46b *Macrocystis* **stipe** in cross section. Specialized cells called "sieve elements," present in all vascularized land plants, have also evolved in some brown algae to transport the photosynthetic products (prep. slide, c.s., 125×). (Photo by J. W. Perry)

Figure 46c *Macrocystis* **stipe** in longitudinal section. See caption for Figure 46b (prep. slide, l.s., 100×). (Photo by J. W. Perry)

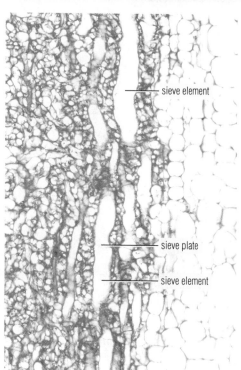

sieve element

sieve plate

sieve element

blade

pneumatocyst

stipe

Figure 46d *Nereocystis*, a Pacific **kelp** with an air-filled vesicle called a **pneumatocyst** (float) (live, 0.3×). (Photo by J. W. Perry)

Div. Chromophyta

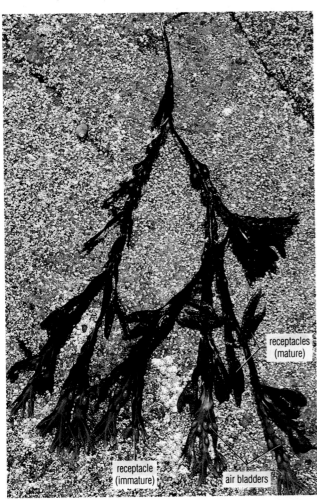

Figure 47a *Fucus* (**rockweed**), a brown alga that grows along both coasts of the United States (live, 0.75×). (Photo by J. W. Perry)

Figure 47b *Fucus* colony. The swollen receptacles at the ends of the branches contain the sex organs. "Dots" on the receptacles are openings to **conceptacles**, shown in Figures 47c and 47d (live, 0.35×). (Photo by J. W. Perry)

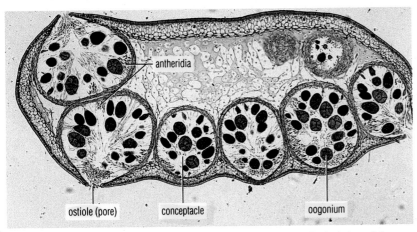

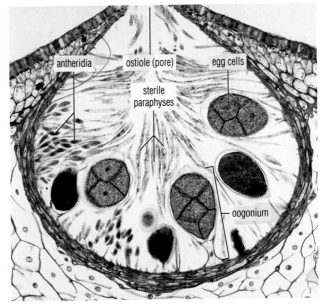

Figure 47c *Fucus* **receptacle**. The cavities (conceptacles) are the location of the sex organs (prep. slide, c.s., 45×). (Photo by J. W. Perry)

Figure 47d *Fucus* **conceptacle** containing male **antheridia** and female **oogonia**. This is a monoecious species, having both male and female gametangia in the same conceptacle, and is found along the Atlantic Coast. Dioecious species (with separate male and female conceptacles on separate plants) are found along the Pacific Coast (prep. slide, c.s., 110×). (Photo by J. W. Perry)

Figure 48a   Transmission electron micrograph of the motile unicell *Chlamydomonas* (t.e.m., l.s., 10,500×). (Photo courtesy H. Hoops)

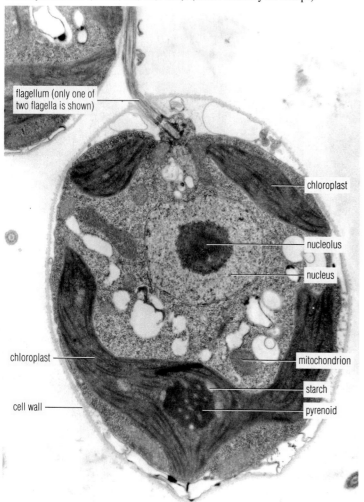

flagellum (only one of two flagella is shown)

chloroplast

nucleolus

nucleus

chloroplast

mitochondrion

starch

cell wall

pyrenoid

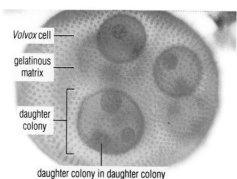

Volvox cell

gelatinous matrix

daughter colony

daughter colony in daughter colony

Figure 48b   *Volvox* **colony** with asexually produced **daughter colonies** (autocolonies) (prep. slide, w.m., 10×). (Photo by J. W. Perry)

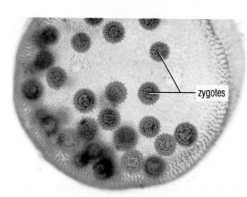

zygotes

Figure 48c   *Volvox* **colony** with sexually produced **zygotes** (prep. slide, w.m., 160×). (Photo by J. W. Perry)

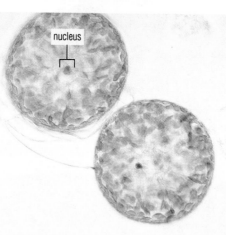

nucleus

Figure 48d   Very large nonmotile unicells, *Eremosphaera*. These cells are so large they can be seen with the unaided eye (live, w.m., 240×). (Photo by J. W. Perry)

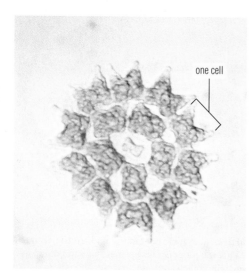

one cell

Figure 48e   *Pediastrum*, a nonmotile colony (live, w.m., 630×). (Photo by J. W. Perry)

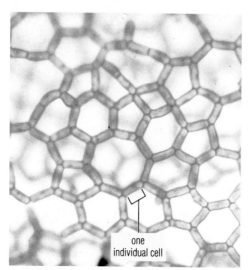

one individual cell

Figure 48f   The **water net**, *Hydrodictyon*, a nonmotile colony (prep. slide, w.m., 250×). (Photo by J. W. Perry)

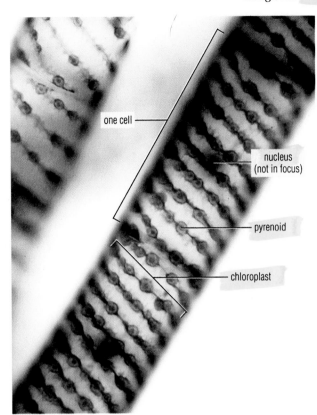

one cell

nucleus (not in focus)

pyrenoid

chloroplast

Figure 49a *Spirogyra*, commonly called **pond scum**, a non-motile filament. The focus is on the spiral **chloroplast** that has enlarged **pyrenoids**, centers for starch synthesis (prep. slide, w.m., 565×). (Photo by J. W. Perry)

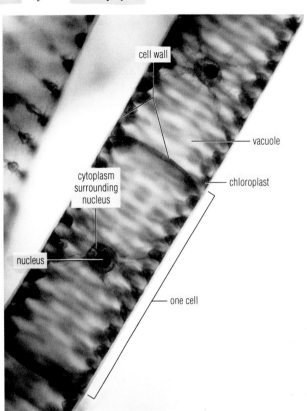

cell wall

vacuole

cytoplasm surrounding nucleus

chloroplast

nucleus

one cell

Figure 49b *Spirogyra*, with the focus on the **nucleus** (prep. slide, w.m., 565×). (Photo by J. W. Perry)

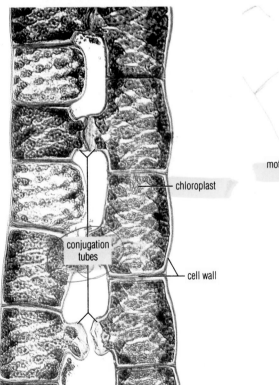

chloroplast

conjugation tubes

cell wall

Figure 49c *Spirogyra* undergoing sexual reproduction by **conjugation**. Conjugation tubes are meeting (prep. slide, w.m., 275×). (Photo by Biodisc, Inc.)

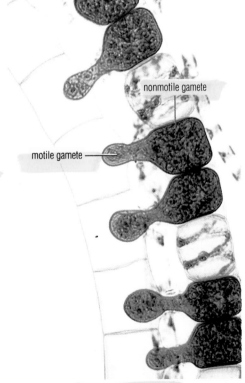

nonmotile gamete

motile gamete

Figure 49d *Spirogyra* **conjugation**. The motile gamete from the left filament is migrating through the conjugation tube and fusing with the nonmotile gamete (prep. slide, w.m., 310×). (Photo by Biodisc, Inc.)

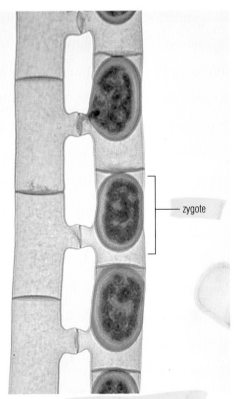

zygote

Figure 49e *Spirogyra* at the end of **conjugation**. Thick-walled **zygotes** have formed in one filament (prep. slide, w.m., 275×). (Photo by Biodisc, Inc.)

Figure 50a
Zygnema, a nonmotile filament closely related to *Spirogyra* but with two star-shaped **chloroplasts** per cell (live, w.m., 250×). (Photo by J. W. Perry)

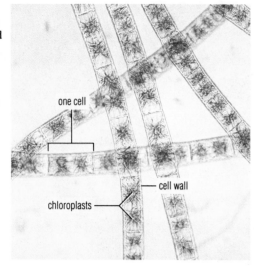

one cell

cell wall

chloroplasts

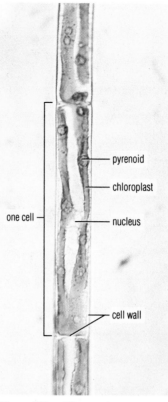

one cell

pyrenoid

chloroplast

nucleus

cell wall

Figure 50b *Mougeotia*, a non-motile filament with a ribbon-shaped chloroplast (live, w.m., 450×). (Photo by J. W. Perry)

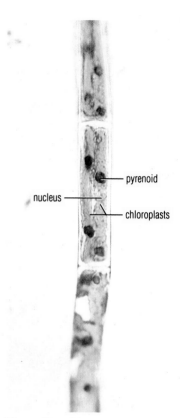

nucleus

pyrenoid

chloroplasts

Figure 50c *Mougeotia* stained with potassium iodide (I_2KI) solution, which stains the starch-containing **pyrenoids** (live, w.m., 280×). (Photo by J. W. Perry)

Figure 50d
Closterium, a **desmid**. **Brownian motion** of the crystals in the terminal vacuole is especially evident in living preparations (live, w.m., 250×). (Photo by J. W. Perry)

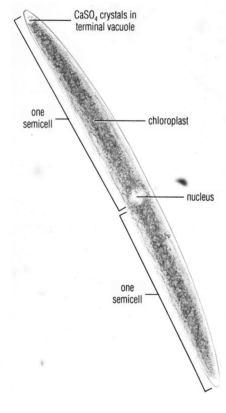

$CaSO_4$ crystals in terminal vacuole

one semicell

chloroplast

nucleus

one semicell

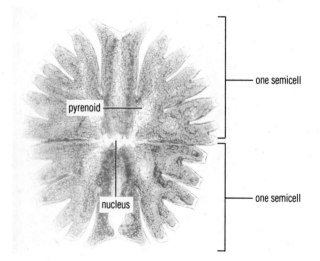

one semicell

pyrenoid

nucleus

one semicell

Figure 50f *Micrasterias*, a **desmid** (live, w.m., 210×). (Photo by J. W. Perry)

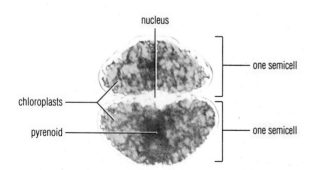

nucleus

one semicell

chloroplasts

pyrenoid

one semicell

Figure 50e *Cosmarium*, a **desmid** (live, w.m., 750×). (Photo by J. W. Perry)

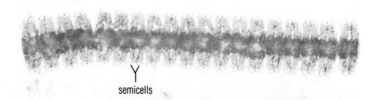

semicells

Figure 50g *Desmidium*. This **desmid** forms a chain because the cells do not separate following cell division (live, w.m., 375×). (Photo by J. W. Perry)

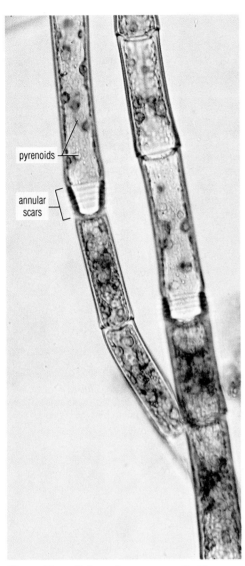

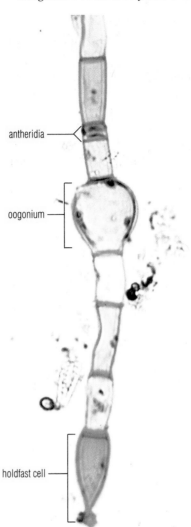

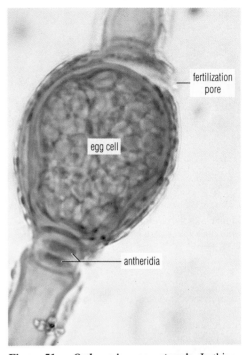

Figure 51c *Oedogonium* **gametangia**. In this species, the sperm-containing **antheridia** are immediately adjacent to the **oogonium**. The pore through which the sperm enter the oogonium is visible (prep. slide, w.m., 800×). (Photo by J. W. Perry)

Figure 51a *Oedogonium*, a nonmotile filament. **Annular scars** are produced by successive cell divisions that occur within the filament (live, w.m., 540×). (Photo by J. W. Perry)

Figure 51b *Oedogonium* filament that has gametangia and a **holdfast cell** that anchors the filament to the substrate (prep. slide, w.m., 400×). (Photo by Biodisc, Inc.)

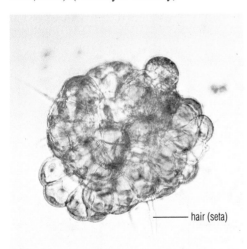

Figure 51d *Coleochaete*, the alga believed to resemble most closely the ancestral alga that gave rise to land plants (live, w.m., 540×). (Photo by J. W. Perry)

Figure 51e *Cladophora* in a tidepool along the Atlantic Coast. Some *Cladophora* species are found in fresh water (live, 0.1×). (Photo by J. W. Perry)

Figure 51f *Cladophora* filament. The branching is characteristic of *Cladophora* (live, w.m., 110×). (Photo by J. W. Perry)

Figure 52a *Codium*, **sea rope**, a marine siphonous green alga (live, 0.5×). (Photo by J. W. Perry)

Figure 52b *Ulva*, **sea lettuce**, a marine tissuelike green alga (live, 0.4×). (Photo by J. W. Perry)

Figure 52c *Chara*, a **stonewort,** with sex organs (live, 2×). (Photo by J. W. Perry)

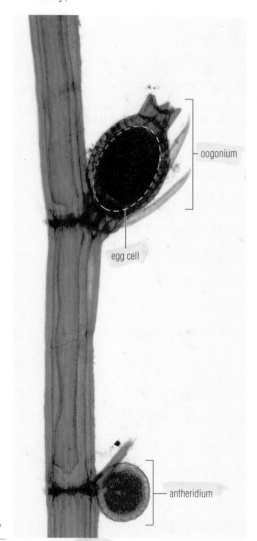

oogonium

egg cell

antheridium

Figure 52d *Chara* gametangia (prep. slide, w.m., 30×). (Photo by J. W. Perry)

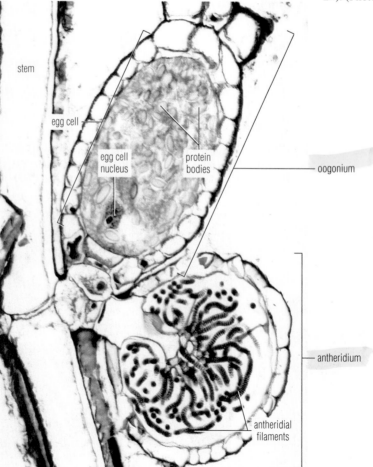

stem

egg cell

egg cell nucleus

protein bodies

oogonium

antheridium

antheridial filaments

Figure 52e *Chara* **gametangia**. In this individual, the gametangia are adjacent to one another. The single **egg cell** fills the **oogonium** (prep. slide, l.s., 120×). (Photo by J. W. Perry)

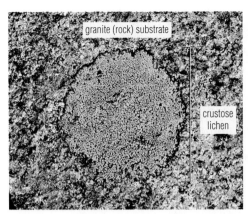

Figure 53a **Crustose lichen** growing on rock. Lichens are **pioneer species**, being the first to colonize bare rock (live, 0.5×). (Photo by J. W. Perry)

Figure 53b **Crustose lichen** with ascospore-containing **apothecia** (live, 0.5×). (Photo by J. W. Perry)

Figure 53c **Foliose lichen** (*Parmelia sulcata*) growing on the bark of a white birch tree (live, 0.5×). (Photo by J. W. Perry)

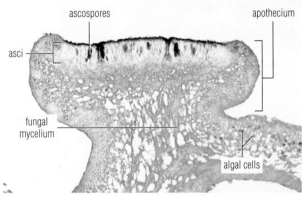

Figure 53d Section of the **foliose lichen** *Physcia* with spore-bearing **apothecium** (ascocarp) (prep. slide, sec., 84×). (Photo by J. W. Perry)

Figure 53e **Fruticose lichen** (*Usnea*) on bald cypress tree sapling (live, 0.3×). (Photo by J. W. Perry)

Figure 53f **Fruticose lichen**. The cups at the tips of the branches are ascospore-containing **apothecia** (live, 1×). (Photo by J. W. Perry)

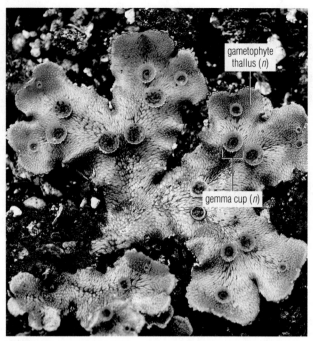

Figure 54a **Liverwort gametophyte thalli** (*Ricciocarpus*) (live, 1.5×). (Photo by J. W. Perry)

Figure 54b **Liverwort** (*Marchantia*) **thallus** with **gemma cups** containing gemmae, which can grow into new thalli (live, 2×). (Photo by J. W. Perry)

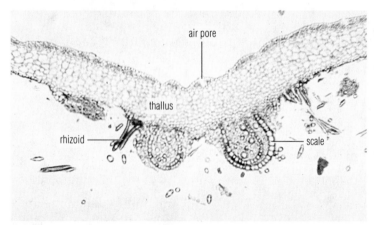

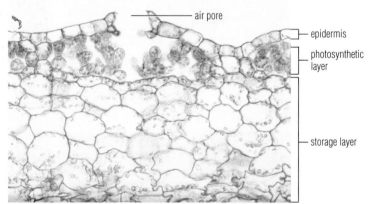

Figure 54c **Liverwort** (*Marchantia*) **thallus**. All structures are gametophytic and haploid (*n*) (prep. slide, c.s., 25×). (Photo by J. W. Perry)

Figure 54d **Liverwort** (*Marchantia*) **thallus** with **air pore** (prep. slide, c.s., 250×). (Photo by J. W. Perry)

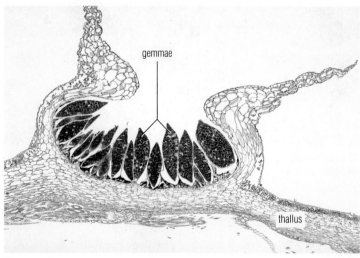

Figure 54e **Liverwort** (*Marchantia*) **thallus** with gemma cup (prep. slide, c.s., 25×). (Photo by J. W. Perry)

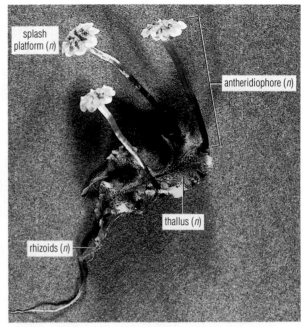

Figure 55a Colony of **liverwort** (*Marchantia*) **male thalli** with **antheridiophores** (live, 0.7×). (Photo by J. W. Perry)

Figure 55b Liverwort (*Marchantia*) **male thallus** with **antheridiophores**. The umbrella-like platform at the top of the antheridiophore is sometimes called a "splash platform" (live, 1×). (Photo by J. W. Perry)

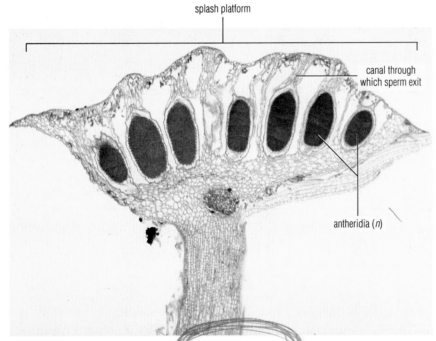

Figure 55c Liverwort (*Marchantia*) **antheridiophore** (prep. slide, l.s., 30×). (Photo by J. W. Perry)

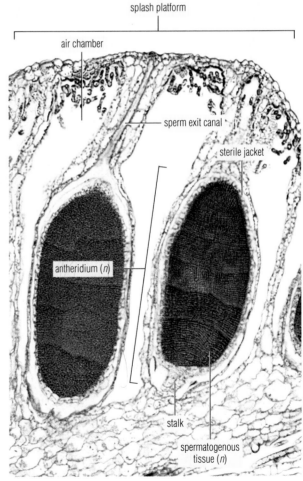

Figure 55d Liverwort (*Marchantia*) **antheridia** (prep. slide, l.s., 120×). (Photo by J. W. Perry)

Figure 56a Liverwort (*Marchantia*) **female thalli** with **archegoniophores** (live, 1.3×). (Photo by J. W. Perry)

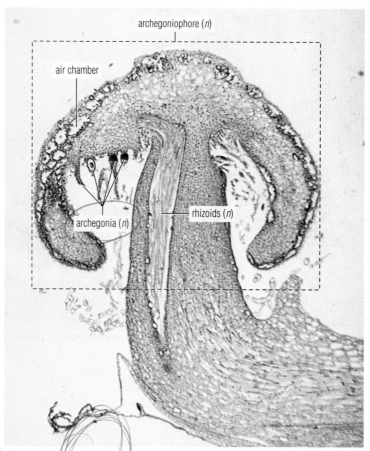

Figure 56b Liverwort (*Marchantia*) **archegoniophore** (prep. slide, l.s., 25×). (Photo by J. W. Perry)

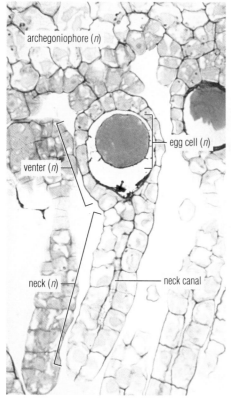

Figure 56c Liverwort (*Marchantia*) **archegonium** (prep. slide, l.s., 380×). (Photo by J. W. Perry)

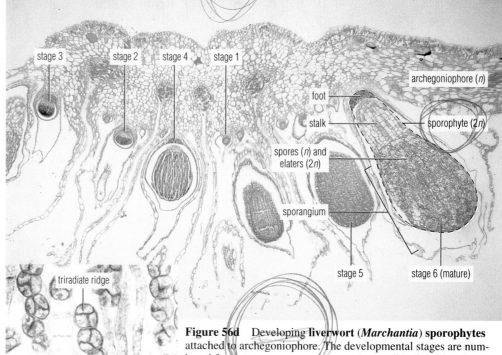

Figure 56d Developing **liverwort** (*Marchantia*) **sporophytes** attached to archegoniophore. The developmental stages are numbered from youngest to most mature (prep. slide, l.s., 30×). Inset at left shows a high magnification of the **spore tetrads**. The triradiate ridge on each spore tetrad is a marker indicating that meiosis has occurred (prep. slide, w.m., 400×). (Photos by J. W. Perry)

Figure 57a Hornwort (*Anthoceros*) **gametophyte thalli** with hornlike **sporophytes** (live, 2.5×). (Photo by J. W. Perry)

sporophytes (2*n*)

cyanobacteria on soil surface

gametophyte thalli (*n*)

sporangium (capsule) (2*n*)

sporophyte (2*n*)

gametophytes (*n*)

Figure 57b *Sphagnum* **gametophytes** with **sporophytes** (live, 1×). (Photo by M. Dibben)

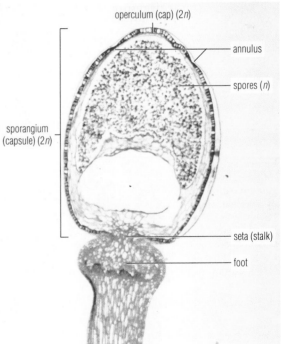

operculum (cap) (2*n*)

annulus

spores (*n*)

sporangium (capsule) (2*n*)

seta (stalk)

foot

Figure 57c *Sphagnum* **sporophyte** (prep. slide, l.s., 28×). (Photo by J. W. Perry)

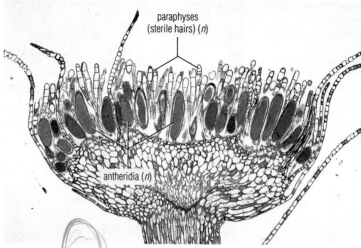

Figure 58a Moss (*Polytrichum*) **gametophytes**. Antheridia are embedded in the cups atop the male gametophytes (live, 0.5×). (Photo by J. W. Perry)

Figure 58b Moss (*Mnium*) **antheridial head** (prep. slide, l.s., 24×). (Photo by J. W. Perry)

Figure 58c Moss (*Mnium*) **antheridia** (prep. slide, l.s., 98×). (Photo by J. W. Perry)

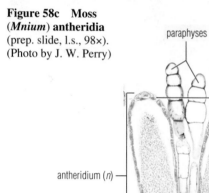

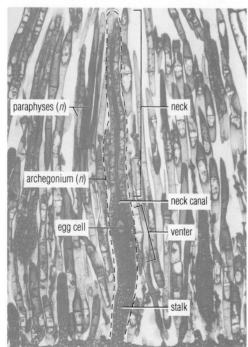

Figure 58d Moss (*Mnium*) **archegonial head** (prep. slide, l.s., 25×). (Photo by J. W. Perry)

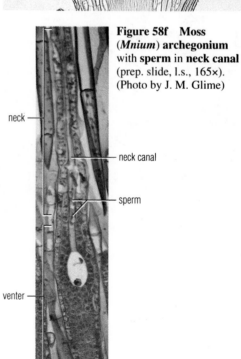

Figure 58e Moss (*Mnium*) **archegonium** (surrounded by a dashed line). **Paraphyses** are believed to aid in water retention (prep. slide, l.s., 90×). (Photo by J. W. Perry)

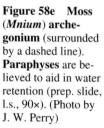

Figure 58f Moss (*Mnium*) **archegonium** with **sperm** in **neck canal** (prep. slide, l.s., 165×). (Photo by J. M. Glime)

Figure 59a   Moss (*Polytrichum*) **sporophytes** growing from **female gametophytes** (live, 0.5×). (Photo by J. W. Perry)

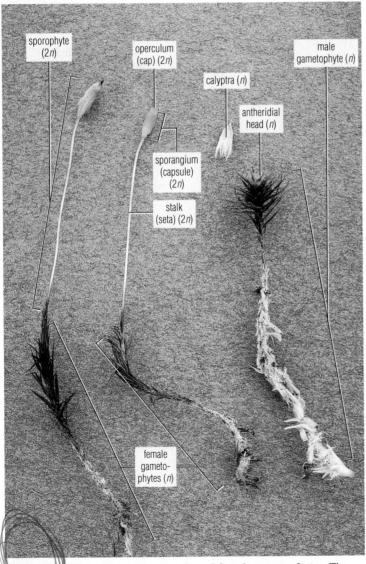

Figure 59b   Moss (*Polytrichum*) **male and female gametophytes**. The female gametophytes have attached **sporophytes**. The **calyptra** has been removed from the sporangium on the right (live, 1.2×). (Photo by J. W. Perry)

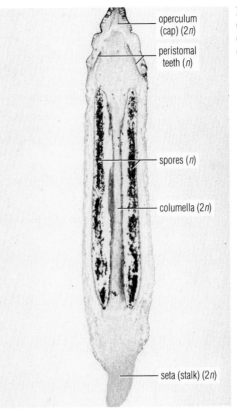

Figure 59c   Moss (*Mnium*) **sporangium** (prep. slide, l.s., 24×). (Photo by J. W. Perry)

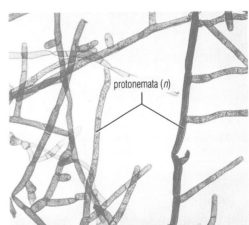

Figure 59d Moss (*Mnium*) **protonemata** that have been grown from moss spores (prep. slide, w.m., 80×). (Photo by J. W. Perry)

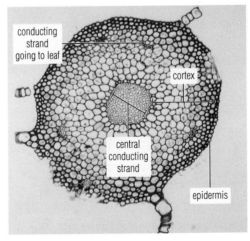

Figure 59e   Moss (*Polytrichum*) **stem**. Though not considered to have true vascular tissue, some mosses have tissue that serves the same conductive function (prep. slide, c.s., 20×). (Photo by J. W. Perry)

Figure 60a **Whisk fern (*Psilotum nudum*) sporophyte**. Note dichotomous branching of this aerial stem (live, 0.75×). (Photo by J. W. Perry)

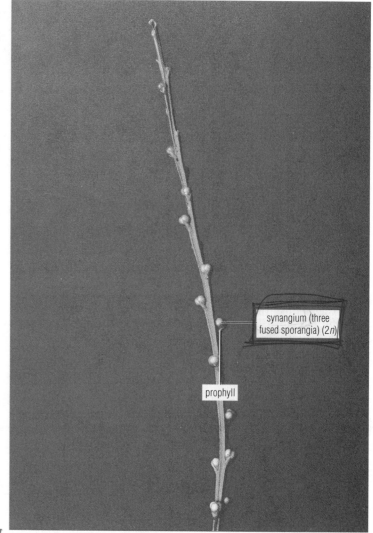

Figure 60b **Whisk fern (*Psilotum nudum*)** aerial shoot with **synangia** (fused sporangia) (live, 1×). (Photo by J. W. Perry)

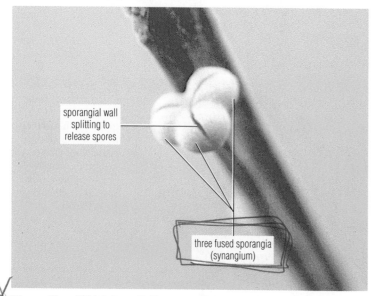

Figure 60c **Whisk fern (*Psilotum nudum*) synangium** at time of spore release (live, 9×). (Photo by J. W. Perry)

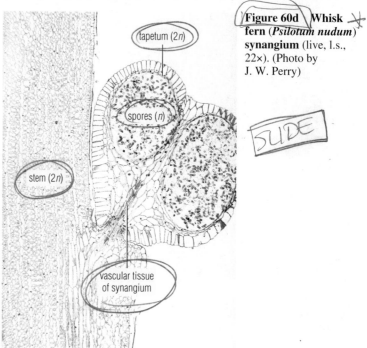

Figure 60d **Whisk fern (*Psilotum nudum*) synangium** (live, l.s., 22×). (Photo by J. W. Perry)

SLIDE

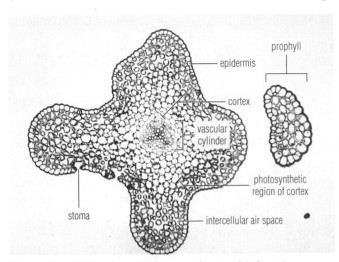

Figure 61a   Whisk fern (*Psilotum nudum*) aerial shoot (prep. slide, c.s., 35×). (Photo by J. W. Perry)

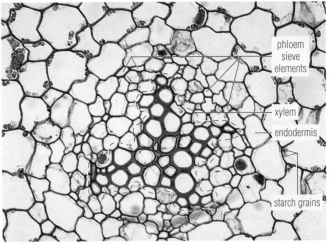

Figure 61b   Whisk fern (*Psilotum nudum*) vascular tissue. This arrangement of vascular tissue with a solid core of xylem is called a **protostele** (prep. slide, c.s., 85×). (Photo by J. W. Perry)

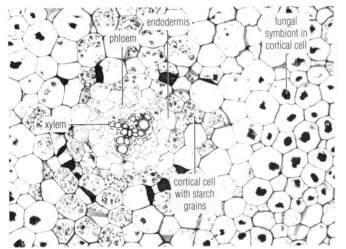

Figure 61c   Whisk fern (*Psilotum nudum*) rhizome, with fungal symbiont in cortex (prep. slide, c.s., 50×). (Photo by J. W. Perry)

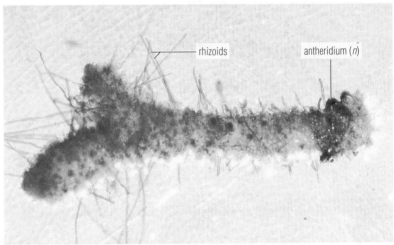

Figure 61d   Whisk fern (*Psilotum nudum*) gametophyte (live, 17×). (Photo by J. W. Perry)

Figure 61e   *Tmesipteris* aerial shoots. *Tmesipteris* is restricted to New Zealand, Australia, and several South Pacific islands (live, c.s., 1×). (Photo by J. W. Perry)

gemmae (2*n*)

gemma holder

sporangium (2*n*)

sporophyll (2*n*)

Figure 62a   **Club moss,** *Huperzia* (*Lycopodium*) *lucidulum*, **sporophyte**. Sporangia in this species are not organized into distinct cones (strobili) (live, 1.6×). Inset shows tip of plant with gemmae (1.6×). (Photo by J. W. Perry)

cone (strobilus) (2*n*)

Figure 62b   **Club moss (***Lycopodium obscurum***) sporophyte**. The sporangia are aggregated in **cones** (**strobili**) at the top of the plant (live, 0.8×). (Photo by J. W. Perry)

cones (strobili) (2*n*)

Figure 62c   **Club moss,** *Lycopodium lagopus* (*complanatum*), **sporophyte**. The cones are elevated on their own specialized branch in this species (live, 0.8×). (Photo by J. W. Perry)

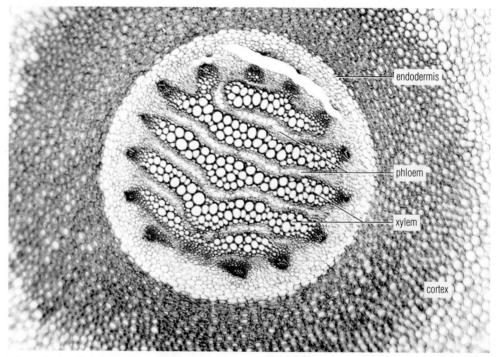

endodermis

phloem

xylem

cortex

Figure 62d   **Club moss (***Lycopodium***) stem**, an example of a **plectostele** (prep. slide, c.s., 100×). (Photo by H. M. Clarke)

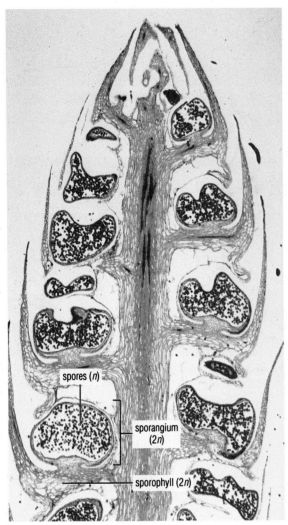

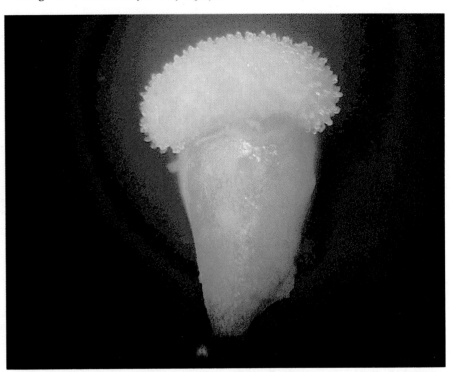

Figure 63b Young **club moss** (*Lycopodium*) **gametophyte**. This gametophyte grows beneath the soil surface. Some club moss species produce green gametophytes that grow on the soil surface (live, 5×). (Photo by Dean P. Whittier)

Figure 63a **Club moss** (*Lycopodium*) **cone** (strobilus) (prep. slide, l.s., 20×). (Photo by J. W. Perry)

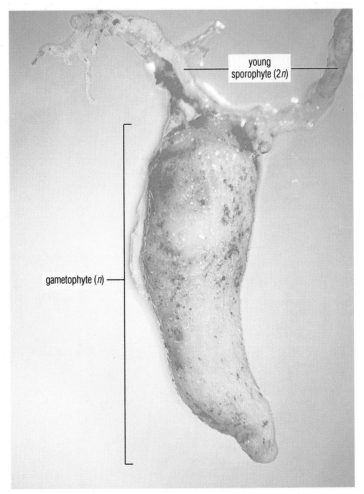

Figure 63c **Club moss** (*Lycopodium*) **gametophyte** with young sporophyte (live, 5×). (Photo by Dean P. Whittier)

Figure 64a Spike moss (*Selaginella paullescens*) **sporophyte** (live, 0.5×). (Photo by J. W. Perry)

Figure 64b Spike moss (*Selaginella*) branch with **cones** (**strobili**). The sporangia of all species of *Selaginella* are organized into cones (live, 1×). (Photo by J. W. Perry)

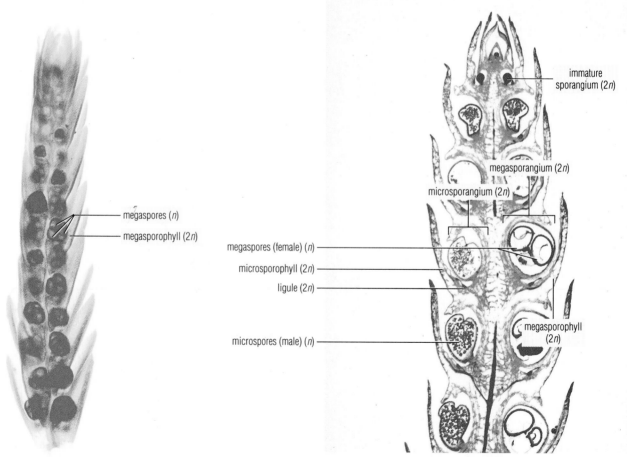

Figure 64c Spike moss (*Selaginella*) **cone** that has been cleared of chlorophyll (prep. slide, w.m., 11×). (Photo by J. W. Perry)

Figure 64d Spike moss (*Selaginella*) **cone**. Note that spike mosses are **heterosporous**; i.e., distinct male spores (**microspores**) and female spores (**megaspores**) are produced (prep. slide, l.s., 18×). (Photo by J. W. Perry)

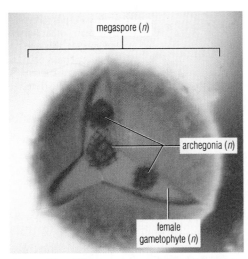

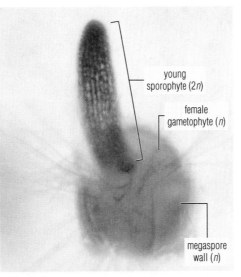

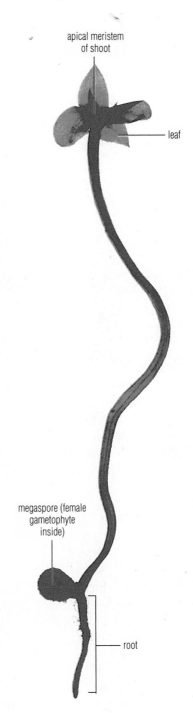

Figure 65a **Megaspore** containing a **female gametophyte** (about one month old) that has produced **archegonia** (live, w.m., 70×). (Photo by W. Carl Taylor)

Figure 65b **Megaspore, female gametophyte**, and young **sporophyte** (about one month old) that has grown from fertilized archegonium (live, w.m., 65×). (Photo by W. Carl Taylor)

Figure 65d Aquatic **quillwort** (*Isoetes*) **sporophytes** in their natural environment (live, 0.5×). (Photo by W. Carl Taylor)

Figure 65e Quillwort (*Isoetes*) sporophyte (live, 0.7×). (Photo by W. Carl Taylor)

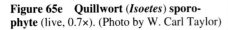

Figure 65c Young spike moss (*Selaginella*) **sporophyte** (sporling) (prep. slide, w.m., 0.8×). (Photo by J. W. Perry)

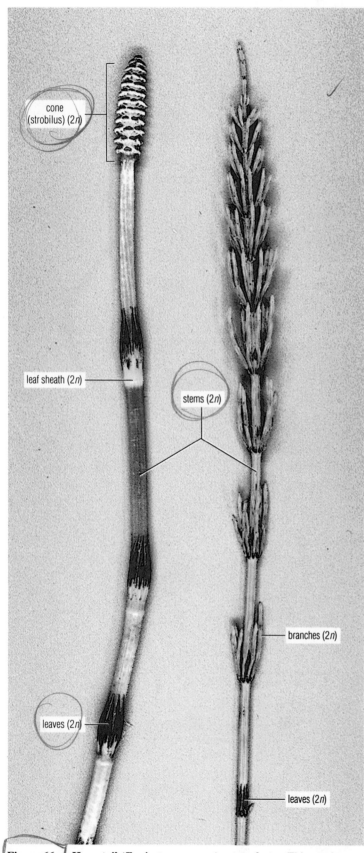

Figure 66b **Horsetail (*Equisetum sylvaticum*) sporophyte**. Like many other horsetail species, this one produces cones on shoots that are also photosynthetic (live, 1.7×). (Photo by J. W. Perry)

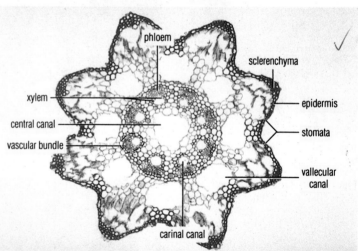

Figure 66c **Horsetail (*Equisetum*) stem**. The characteristic stem anatomy in horsetails may be used to identify different species (prep. slide., c.s., 36×). (Photo by J. W. Perry)

Figure 66a **Horsetail (*Equisetum arvense*) sporophytes**. This species produces two types of shoots from the same rhizome. The one on the left comes up early in the spring, has very little chlorophyll, and produces a spore-containing **cone** (**strobilus**) at its tip. The sterile shoot on the right comes up several weeks later, does most of the plant's photosynthesis, and never produces cones (live, 1.7×). (Photo by J. W. Perry)

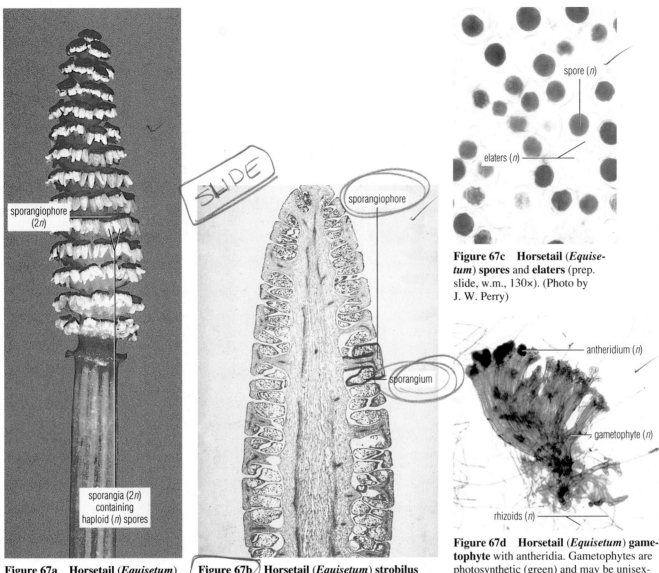

Figure 67a Horsetail (*Equisetum*) **strobilus**. These strobili, in contrast to those of club moss and spike moss genera, are made up of much shortened branches called **sporangiophores** (live, 3×). (Photo by J. W. Perry)

Figure 67b **Horsetail (*Equisetum*) strobilus** (prep. slide, l.s., 7×). (Photo by J. W. Perry)

Figure 67c Horsetail (*Equisetum*) **spores** and **elaters** (prep. slide, w.m., 130×). (Photo by J. W. Perry)

Figure 67d Horsetail (*Equisetum*) **gametophyte** with antheridia. Gametophytes are photosynthetic (green) and may be unisexual or bisexual (prep. slide, w.m., 22×). (Photo by J. W. Perry)

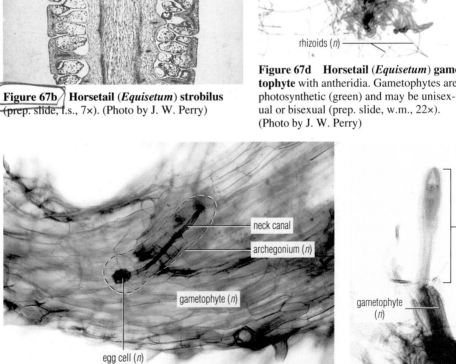

Figure 67e Portion of **horsetail (*Equisetum*) gametophyte** with archegonium (prep. slide, w.m., 130×). (Photo by H.M. Clarke)

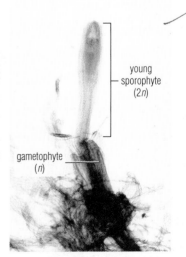

Figure 67f Horsetail (*Equisetum*) **gametophyte** with young sporophyte (prep. slide, w.m., 50×). (Photo by J. W. Perry)

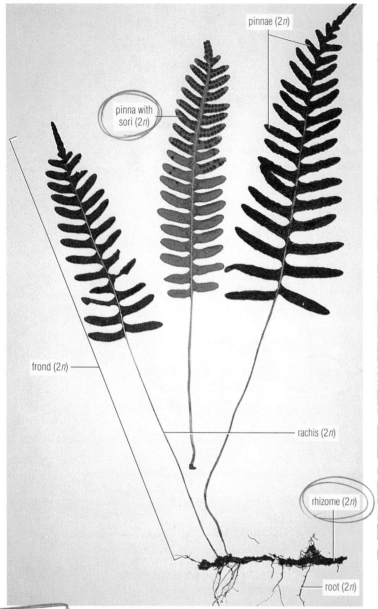

pinnae (2n)

pinna with sori (2n)

frond (2n)

rachis (2n)

rhizome (2n)

root (2n)

Figure 68a Morphology of a **typical fern** (*Polypodium virginianum*). The middle frond (leaf) has its lower surface facing upward, revealing clusters of sporangia called **sori** (herbarium specimen, 1×). (Photo by J. W. Perry)

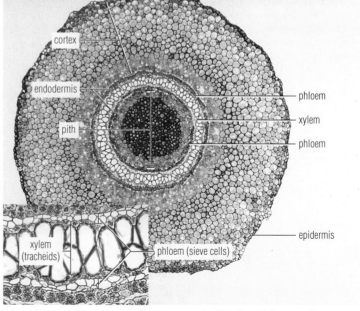

Figure 68b   Fronds of the ostrich fern (*Matteuccia struthiopteris*). Note the tips of the fronds, which are not yet fully elongated, and compare with Figure 68c, a still earlier stage of development of the frond. This fern produces its sporangia on specialized fertile fronds (live, 0.4×). (Photo by J. W. Perry)

fertile frond (2n)

sterile frond (2n)

last year's frond (2n)

Figure 68c Very young **fern frond** just emerging from the soil in early spring. The coiled fronds are called **croziers** or **fiddleheads** (live, 1×). (Photo by J. W. Perry)

pinnae (2n)

cortex

endodermis

pith

phloem

xylem

phloem

epidermis

xylem (tracheids)

phloem (sieve cells)

Figure 68d   Fern rhizome (*Dennstaedtia*). This is a **siphonostele** (prep. slide, c.s., 29×). Inset shows higher magnification of vascular tissue (80×). (Photo by J. W. Perry)

Figure 69a   Tree fern (*Spaeropteris cooperi*), one of the largest living species (live, 0.05×). (Photo by J. W. Perry)

Figure 69b   The aquatic mosquito fern *Azolla*, the smallest living species. This is a common inhabitant of quiet waters of the southern United States. In rice paddies, the cyanobacterial symbiont *Anabaena* (Figure 24b) inhabits the leaves of *Azolla*, capturing atmospheric nitrogen, which is released as the fern dies, fertilizing the rice paddy (live, 2.1×). (Photo by J. W. Perry)

Figure 69c   The water fern *Salvinia*. Unlike most ferns, the water ferns *Azolla* and *Salvinia* are heterosporous (live, 2×). (Photo by J. W. Perry)

Figure 70a Undersurface of **holly fern** (*Cyrtomium falcatum*) **frond**. Each "dot" is a **sorus** (plural, sori), which contains numerous sporangia (live, 0.6×). (Photo by J. W. Perry)

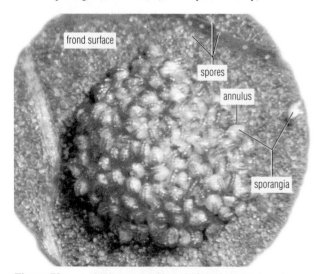

Figure 70c A single **sorus** of **hare's foot fern** (*Polypodium aureum*). The sorus of this fern species lacks an indusium. Each sporangium has a row of cells with thickened cell walls, the **annulus** (live, w.m., 26×). (Photo by J. W. Perry)

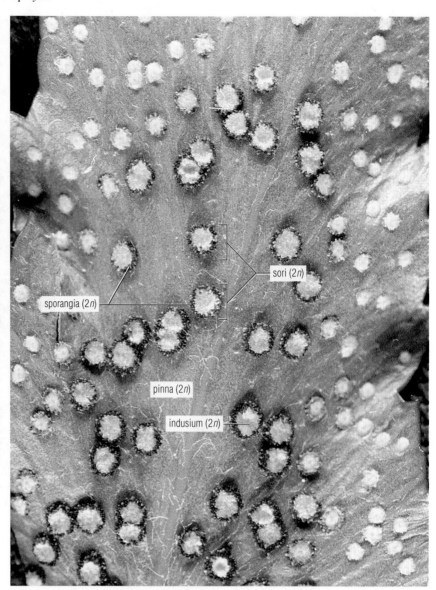

Figure 70b Detail of **holly fern pinna** (leaflet). Brown **sporangia** protrude from the edge of the **indusium**, an umbrella-shaped covering over each sorus (live, 3.8×). (Photo by J. W. Perry)

Figure 70d Section through the **sorus** of **holly fern** (*Cyrtomium*) (prep. slide, c.s., 58×). (Photo by J. W. Perry)

SLIDE

gametophyte (*n*)

archegonia (*n*)

antheridia (*n*)

rhizoids (*n*)

Figure 71a Undersurface of **fern gametophyte**, showing **gametangia** (**antheridia** and **archegonia**). The gametophyte is sometimes called a **pro-thallus** (**prothallium**) (prep. slide, w.m., 22×). (Photo by J. W. Perry)

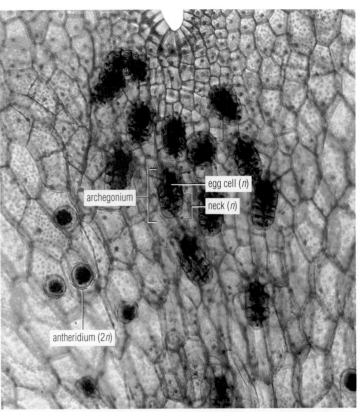

archegonium

egg cell (*n*)

neck (*n*)

antheridium (2*n*)

Figure 71b Gametangia (**antheridia** and **archegonia**) on undersurface of **fern gametophyte** (prep. slide, w.m., 110×). (Photo by J. W. Perry)

Figure 71c Section of a **fern gametophyte** with **antheridia** (prep. slide, l.s., 115×). (Photo by J. W. Perry)

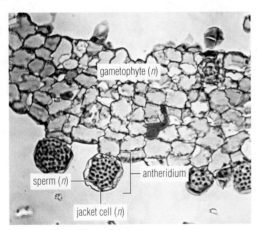

gametophyte (*n*)

sperm (*n*)

antheridium

jacket cell (*n*)

Figure 71d Section of a **fern gametophyte** with **archegonia** (prep. slide, l.s., 230×). (Photo by J. W. Perry)

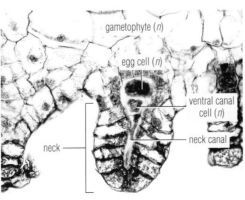

gametophyte (*n*)

egg cell (*n*)

ventral canal cell (*n*)

neck

neck canal

Slide

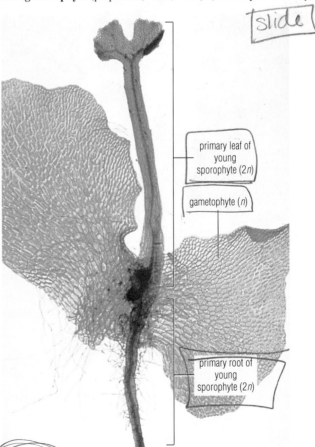

primary leaf of young sporophyte (2*n*)

gametophyte (*n*)

primary root of young sporophyte (2*n*)

Figure 71e **Fern gametophyte** with an attached **sporophyte** produced by fertilization of an egg in one of the archegonia (prep. slide, w.m., 17×). (Photo by Biodisc, Inc.)

Figure 72a Sago palm (*Cycas revoluta*) **male sporophyte** with **pollen cone** (strobilus) (live, 0.04×). (Photo by A. B. Russell)

Figure 72b Sago palm (*Cycas revoluta*) **female sporophyte** (live, 0.04×). (Photo by J. W. Perry)

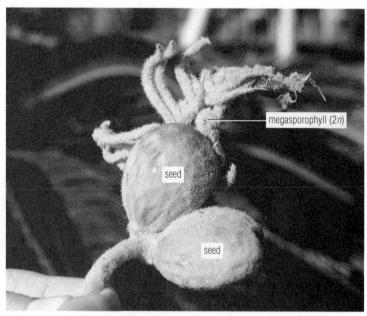

Figure 72c **Megasporophyll** and **seeds** of sago palm. Megasporophylls are not organized in strobili in this species (live, 1×). (Photo by J. W. Perry)

Figure 73a **Male *Zamia* with pollen cones** (strobili). The stem of *Zamia* is, for the most part, beneath the soil line (live, 0.3×). (Photo by J. W. Perry)

Figure 73b **Pollen cone** of *Zamia* (live, 1.2×). (Photo by J. W. Perry)

Figure 73c **Female *Zamia* with seed cones** (live, 0.3×). (Photo by J. W. Perry)

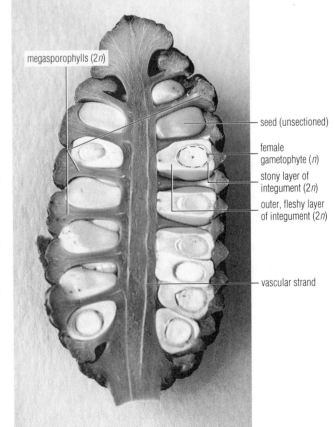

Figure 73d ***Zamia* seed cone** (preserved specimen, l.s., 0.5×). (Photo by J. W. Perry)

Figure 74a **Maidenhair tree**, ***Ginkgo biloba***, the only living species of the phylum. **Sporophyte** (live, 0.01×). (Photo by J. W. Perry)

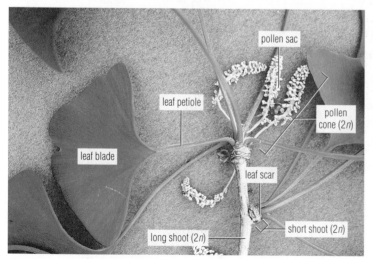

Figure 74c *Ginkgo* **short shoot** with **pollen cones** at pollen shed stage. The short shoot pointed out here is in its third growing season (live, 0.75×). (Photo by J. W. Perry)

Figure 74b A portion of the shoot of *Ginkgo*, showing the very distinctive leaves. Like many gymnosperms, *Ginkgo* produces **long shoots** and **short (spur) shoots**. This short shoot is 7 to 9 years old (live, 0.9×). (Photo by J. W. Perry)

Figure 74d *Ginkgo* **seeds** on a very old (20+ years) short shoot. The fleshy seed coat deteriorates upon ripening, producing a foul odor (some say it smells like "puppy poop"). Consequently, female trees are seldom planted (live, 0.5×). (Photo by J. W. Perry)

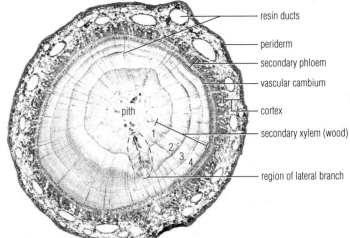

Figure 75b *Pinus* (**pine**) stem that is four years old. Numbers refer to the year when the annual ring was produced in the wood (prep. slide, c.s., 15×). (Photo by J. W. Perry)

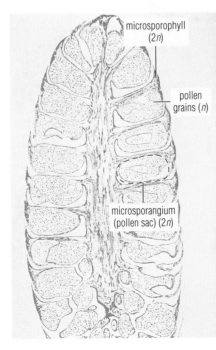

Figure 75a *Pinus* (**pine**) **tree**, the **sporophyte** generation. This is white pine (*P. strobus*), a species prized for its wood (live, 0.01×). (Photo by J. W. Perry)

Figure 75d *Pinus* **pollen** (male) **cone** (strobilus) containing **pollen grains**, which are immature male gametophytes (prep. slide, l.s., 12×). (Photo by J. W. Perry)

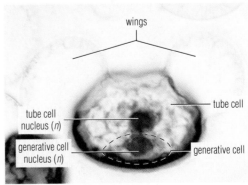

Figure 75c *Pinus* (**pine**), cluster of **pollen** (male) **cones** (strobili) shedding **pollen** (live, 1×). (Photo by J. W. Perry)

Figure 75e *Pinus* **pollen grain**, the immature male gametophyte (section, 880×). (Photo by J. W. Perry)

leaves (needles) (2*n*)

two-month-old
female cones (2*n*)

fourteen-month-old
female cones (2*n*)

Figure 76a *Pinus* **seed** (female) **cones** (strobili). At the tip are tiny female cones about two months old; below them is a cluster of cones in their second year of development (about fourteen months old). This species (red pine, *P. resinosa*) requires three seasons for maturation of the cones before the seeds are shed (live, 0.8×). (Photo by J. W. Perry)

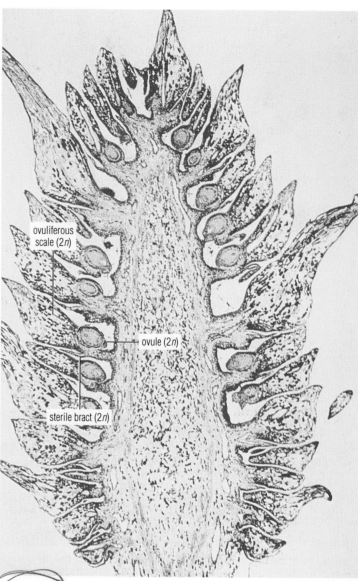

ovuliferous
scale (2*n*)

ovule (2*n*)

sterile bract (2*n*)

Figure 76c *Pinus*, young (two-month-old) **seed** (female) **cone** (prep. slide, l.s., 17×). (Photo by J. W. Perry)

ovuliferous
scale (2*n*)

sterile bract (2*n*)

Figure 76b Douglas fir (*Pseudotsuga*), mature **seed** (female) **cones**. In contrast with those of pine, these cones have very long sterile bracts (live, 0.5×). (Photo by J. W. Perry)

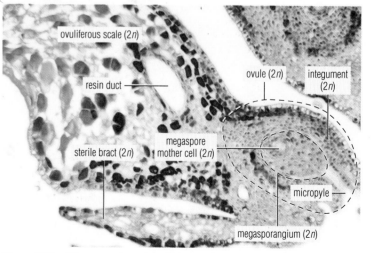

ovuliferous scale (2*n*)

resin duct

sterile bract (2*n*)

megaspore
mother cell (2*n*)

ovule (2*n*)

integument (2*n*)

micropyle

megasporangium (2*n*)

Figure 76d *Pinus*, portion of **seed** (female) **cone** showing **megaspore mother cell** within **megasporangium** (prep. slide, l.s., 93×). (Photo by J. W. Perry)

Figure 77a *Pinus* **ovule** in stage where the **female gameto-phyte** (surrounded by dashed line) contains **archegonia** (prep. slide, l.s., 15×). (Photo by J. W. Perry)

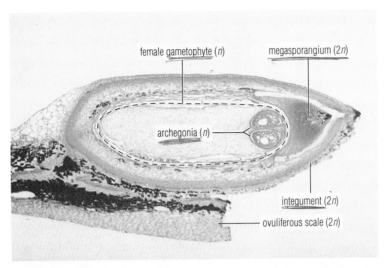

female gametophyte (*n*)   megasporangium (2*n*)

archegonia (*n*)

integument (2*n*)

ovuliferous scale (2*n*)

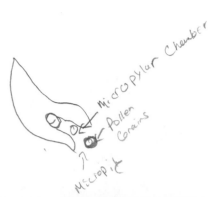

Micropylar Chamber

Pollen Grains

Micropit

Figure 77b *Pinus* **archegonium** contain-ing large **egg cell** at time of **fertilization** (prep. slide, l.s., 97×). (Photo by J. W. Perry)

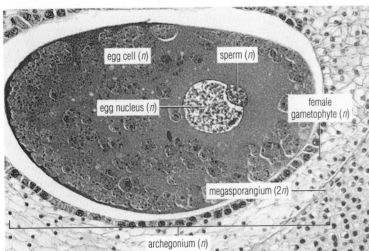

egg cell (*n*)   sperm (*n*)

egg nucleus (*n*)

female gametophyte (*n*)

megasporangium (2*n*)

archegonium (*n*)

winged seed (*n* + 2*n*)

wing of seed

ovuliferous scale (2*n*)

Figure 77c *Pinus*, mature **seed** (female) **cones** and a single **seed** (live, 0.7×). (Photo by J. W. Perry)

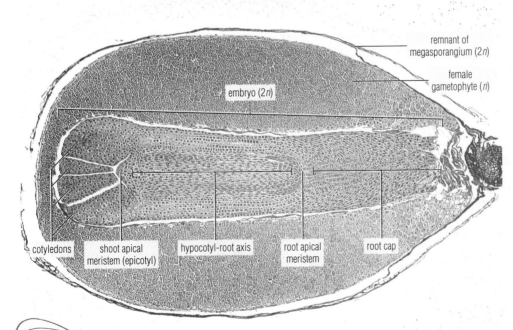

remnant of megasporangium (2*n*)

female gametophyte (*n*)

embryo (2*n*)

cotyledons   shoot apical meristem (epicotyl)   hypocotyl-root axis   root apical meristem   root cap

Figure 77d *Pinus* **seed** containing **embryo**. The hard seed coat was removed during slide preparation (prep. slide, l.s., 31×). (Photo courtesy Biodisc, Inc.)

seed coat (2*n*)

cotyledons (2*n*)

new photosynthetic shoot of sporophyte (2*n*)

hypocotyl (2*n*)

Figure 77e *Pinus* (pinyon pine, *P. edulis*) seedling with **seed** still attached to **cotyledons** (live, 1.6×). (Photo by J. W. Perry)

Figure 78b   Spruce (Norway spruce, *Picea abies*), mature **seed** (female) **cones** (live, 0.2×). (Photo by J. W. Perry)

this year's immature seed cone (2*n*)

last year's seed cone (seeds shed)

Figure 78c   Cedar (white cedar, *Thuja occidentalis*), immature and mature **seed** (female) **cones**. The seed cones of this species mature within a single growing season (live, 0.6×). (Photo by J. W. Perry)

Figure 78a   Spruce (Norway spruce, *Picea abies*) tree with mature **seed** (female) **cones** (live, 0.002×). (Photo by J. W. Perry)

seed cone (2*n*)

Figure 78d   Cedar (red cedar, *Juniperus virginiana*) **seed** (female) **cones**. (This cedar is the alternate host of cedar-apple rust; see Figure 37a) (live, 0.5×). (Photo by J. W. Perry)

aril (2*n*)

Figure 78e   Yew (American yew, *Taxus canadensis*). The seed of this species is enclosed in a fleshy, berrylike structure called an **aril** (live, 0.3×). (Photo by J. W. Perry)

Figure 79a   *Ephedra*. Commonly called Mormon tea or joint fir, this plant grows in the southwestern United States (live, 0.06×). (Photo by J. W. Perry)

Figure 79b   **Male** *Ephedra* (*E. aspera*) plant with pollen cones that look very much like angiosperm flowers (live, 0.2×). (Photo by J. W. Perry)

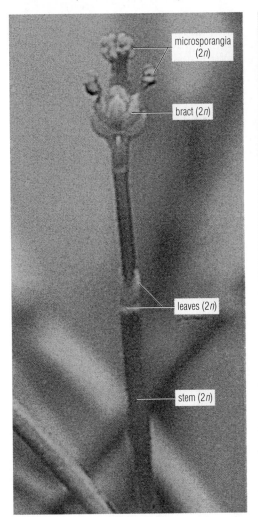

Figure 79c   **Stem morphology** and **pollen cones** of *Ephedra* (live, 2×). (Photo by J. W. Perry)

Figure 79d   **Female** *Ephedra* (*E. aspera*) plant with young **seed cones** (live, 0.6×). (Photo by J. W. Perry)

Figure 79e   Mature **seed cones** of *Ephedra* (live, 1.1×). (Photo by J. W. Perry)

Figure 80a   *Gnetum leyboldii*, a species that grows as a vine. All *Gnetum* species have very angiosperm-like leaves (live, 0.4×). (Photo by J. W. Perry)

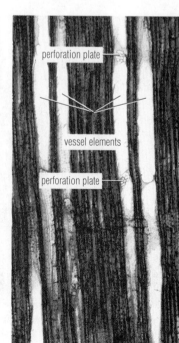

Figure 80b   *Gnetum gnemon*, a species that grows as a tree. This one is only about three years old (live, 0.07×). (Photo by J. W. Perry)

Figure 80c   Pollen cone of *Gnetum* (*G. gnemon*) (live, 1×). (Photo by J. W. Perry)

Figure 80d   **Gnetophyte wood** (*Ephedra*) showing **xylem vessels**, the characteristic that gives the phylum its common name, "vessel-containing gymnoperms" (prep. slide, l.s., 325×). (Photo by J. W. Perry)

Figure 81a *Welwitschia mirabilis*, among the most bizarre plants in the world and restricted to the Namib Desert of southwest Africa. This is a large female plant (live, 0.07×). (Photo by C. H. Bornman)

Figure 81c **Pollen cones** of *Welwitschia mirabilis*. Despite their superficial resemblance to flowers of angiosperms, these are not flowers (live, 1×). (Photo by C. H. Bornman)

Figure 81b Male *Welwitschia mirabilis* showing **stem** and **pollen cones**. The stem is a funnel-shaped structure with a long taproot (live, 0.1×). (Photo by C. H. Bornman)

Figure 81d Female *Welwitschia mirabilis* with **seed cones** (live, 0.25×). (Photo by C. H. Bornman)

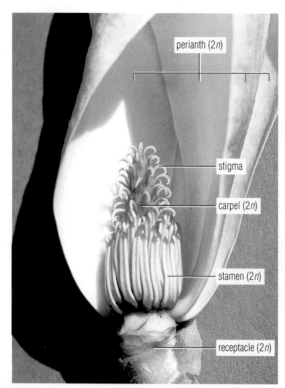

perianth (2n)

stigma

carpel (2n)

stamen (2n)

receptacle (2n)

pistil

stamen

petal

pedicel

bud

sepal

Figure 82a *Magnolia* **flower** (saucer magnolia, *M.* × *Soulanginana*, a hybrid of *M. hepatapeta* and *M. quinquepet*; *Magnoliaceae*), an example of a primitive dicotyledon. This flower lacks distinction between sepals and petals (consequently, the term **tepal** may be applied to these parts), has its floral parts arranged in a spiral pattern on the stem, and has numerous stamens and carpels (pistils) (live, 1×). (Photo by J. W. Perry)

Figure 82b **Dicot flowers** (spring beauty, *Claytonia virginica*; *Portulacaceae*). Dicotyledon flowers typically have their parts in fours or fives. These are regular (radially symmetrical) flowers (live, 1.2×). (Photo by J. W. Perry)

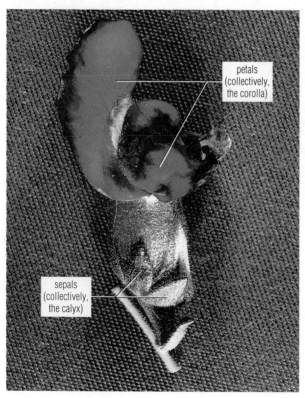

petals (collectively, the corolla)

sepals (collectively, the calyx)

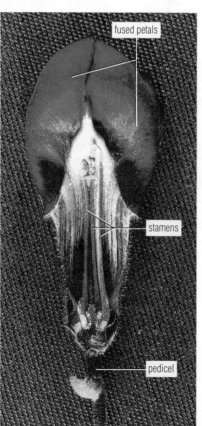

fused petals

stamens

pedicel

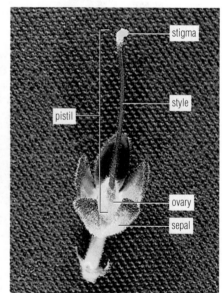

stigma

style

pistil

ovary

sepal

Figure 82e **Dicot flower** (snapdragon), with corolla and stamens removed (live, 2.75×). (Photo by J. W. Perry)

Figure 82c **Dicot flower** (snapdragon, *Antirrhinum majus*; *Scrophulariaceae*), an irregular (bilaterally symmetrical) flower (live, 1.8×). (Photo by J. W. Perry)

Figure 82d **Dicot flower** (snapdragon), with one of the two fused petals and the pistil removed (live, 2×). (Photo by J. W. Perry)

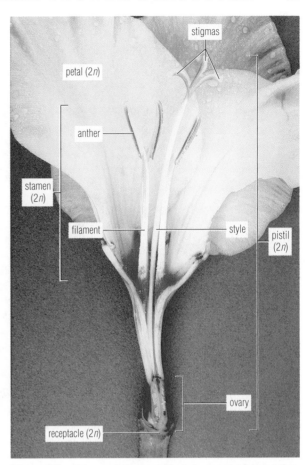

Figure 83a   Monocot flower (wake robin, *Trillium erectum*; *Liliaceae*). Typical of monocotyledons, the floral parts are in threes (live, 1.2×). (Photo by J. W. Perry)

Figure 83b   Monocot flower (*Gladiolus* sp.; *Iridaceae*) that has been sectioned (live, 1×). (Photo by J. W. Perry)

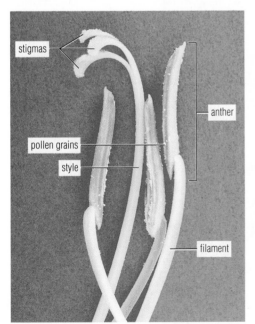

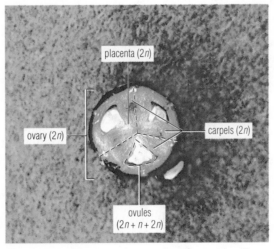

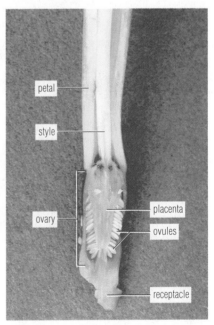

Figure 83d   Monocot flower (*Gladiolus*) **ovary** (live, c.s., 2.75×). (Photo by J. W. Perry)

Figure 83c   Monocot flower (*Gladiolus*), upper portion of **stamens** and **pistil** (live, 1.5×). (Photo by J. W. Perry)

Figure 83e   Monocot flower (*Gladiolus*), lower portion of **pistil** (live, l.s., 1.3×). (Photo by J. W. Perry)

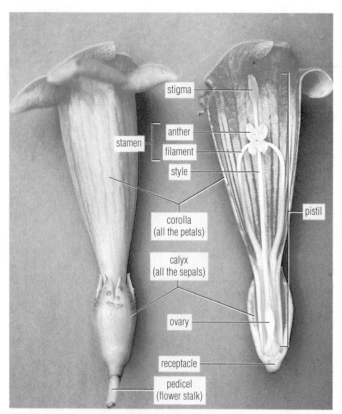

Figure 84a **Hypogynous dicot flower** (trumpet creeper, *Campsis radicans*; *Bignoniaceae*). All floral parts clearly originate below the **superior ovary**. Both the five sepals and the five petals are fused in this **regular**, **complete flower** (live, 1×). (Photo by J. W. Perry)

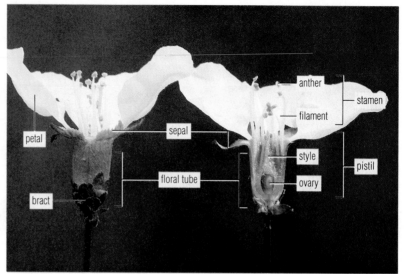

Figure 84b **Perigynous dicot flower** (cherry, *Prunus*; *Rosaceae*). The floral parts are fused together, forming a floral tube. Hence, the petals and stamens appear to come off at the top of this floral tube, which surrounds the **superior ovary**. The flower is **regular and complete**. Its ovary has five separate styles, an indication that the pistil consists of five carpels. (live, 1.4×). (Photo by J. W. Perry)

Figure 84c **Epigynous dicot flower** (*Fuchsia*; *Onagraceae*) (live, 1×). (Photo by J. W. Perry)

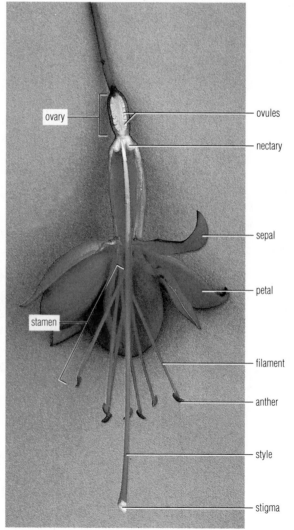

Figure 84d **Epigynous dicot flower** (*Fuchsia*) that has been sectioned (live, l.s., 1.25×). (Photo by J. W. Perry)

Figure 85a **Monocot inflorescence** (cluster of flowers) of a grass (timothy, *Phleum pratense*; *Graminae*). Anthers are abundant at this stage of development (live, 0.5×). (Photo by J. W. Perry)

Figure 85b **Male flowers** of oak (*Quercus* sp.; *Fagaceae*), arranged in an elongated, deciduous inflorescence called a **catkin**. The individual flowers lack petals (live, 1×). (Photo by J. W. Perry)

Figure 85c **Male flower** of begonia (*Begonia* sp.; *Begoniaceae*) (live, 3×). (Photo by J. W. Perry)

Figure 85d **Female flower** of begonia (*Begonia* sp). The perianth parts are not distinguished as sepals and petals, and thus are called **tepals**. The ovary is inferior and "winged" (live, 3×). (Photo by J. W. Perry)

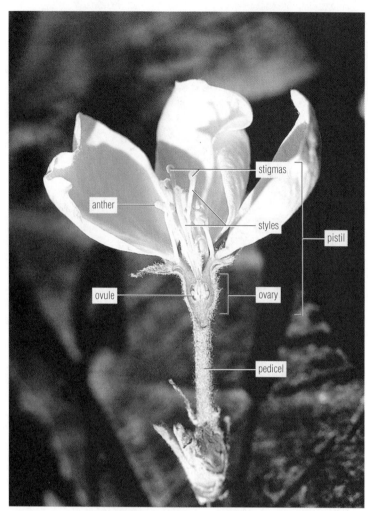

Figure 86a   **Apple flower** (*Pyrus malus*; *Rosaceae*), an epigynous dicot. The floral parts come off above the **inferior ovary**. The flower is **regular and complete** (live, 2.3×). (Photo by J. W. Perry)

Figure 86d   **Strawberry** (*Fragaria ananassa*; *Rosaceae*) **flower** (live, 3×). (Photo by J. W. Perry)

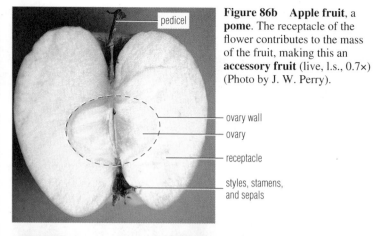

Figure 86b   **Apple fruit**, a **pome**. The receptacle of the flower contributes to the mass of the fruit, making this an **accessory fruit** (live, l.s., 0.7×) (Photo by J. W. Perry).

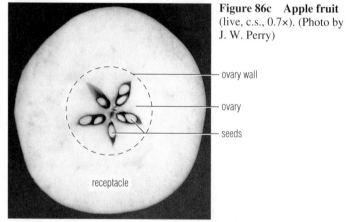

Figure 86c   **Apple fruit** (live, c.s., 0.7×). (Photo by J. W. Perry)

Figure 86e   Immature **strawberry**. The fleshy portion of the strawberry is derived from the flower's receptacle; the actual fruits are the individual seed-like structures, which are **achenes** (live, 2×). (Photo by J. W. Perry)

Figure 86f   Mature **strawberry** with mature **achenes** (live, 0.8×). (Photo by J. W. Perry)

Figure 87a   Raspberry (*Rubus* sp.; *Rosaceae*) **flowers** (live, 1.2×). (Photo by J. W. Perry)

Figure 87b   Black **raspberry fruits**. These are **aggregate fruits** consisting of numerous small subunits called **fruitlets**. Each fruitlet is a **drupe** (live, 0.8×). (Photo by J. W. Perry)

Figure 87c   Banana (*Musa paradisiaca*; *Musaceae*) **plant** with inflorescence and immature fruits, which are **berries**. Banana inflorescences exhibit sexual dimorphism: female flowers at the top develop into fruits without fertilization. In the middle are flowers with male and female parts. Male flowers terminate the floral stalk (live, 0.13×). (Photo by J. W. Perry)

Figure 87d   Female banana flower (live, 0.75×). (Photo by J. W. Perry)

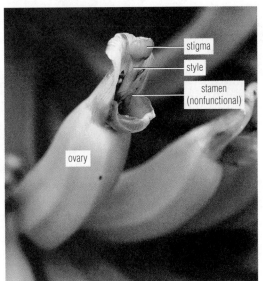

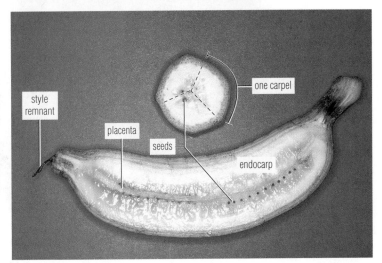

Figure 87e   Banana fruit, a **berry** (live, c.s., l.s., 0.5×). (Photo by J. W. Perry)

Figure 88a **Male flowers** of ash-leaved **maple** (box elder, *Acer negundo*; *Aceraceae*). Box elder flowers lack petals (live, 1×). (Photo by J. W. Perry)

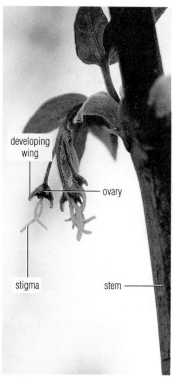

Figure 88b **Female flowers** of ash-leaved **maple** (live, 1×). (Photo by J. W. Perry)

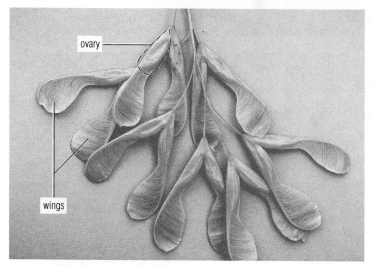

Figure 88c Immature **fruits** (paired **samaras**) of ash-leaved **maple** (live, 1×). (Photo by J. W. Perry)

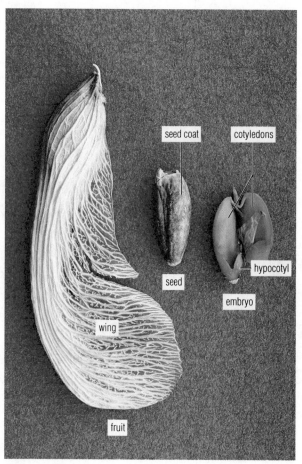

Figure 88d Mature **samara** (**fruit**), **seed**, and **embryo** of silver **maple** (*Acer saccharinum*; *Aceraceae*) (live, 2.1×). (Photo by J. W. Perry)

Figure 89a   **Composite inflorescence** (cluster of flowers) called a **head**. This sunflower (*Helianthus annuus*; *Compositae*) consists of **ray flowers** and **disk flowers** (live, 0.1×). (Photo by J. W. Perry)

Figure 89c   Nearly mature **head** of sunflower (live, 0.5×). (Photo by J. W. Perry)

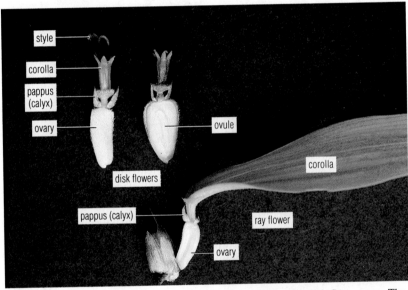

Figure 89b   Two **disk flowers** and a **ray flower** from a sunflower inflorescence. The ovary of the right-hand disk flower is sectioned, showing the single ovule (live, 1.1×). (Photo by J. W. Perry)

Figure 89d   One-seeded sunflower **fruits** called **achenes**. The achene on the left has been cut open to reveal the **seed** (live, 2.3×). (Photo by J. W. Perry)

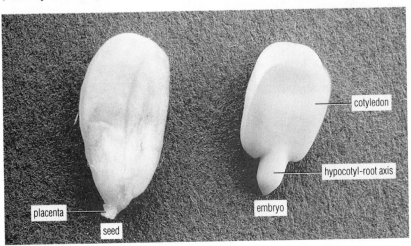

Figure 89e   Sunflower **seed** and **embryo** (live, 5×). (Photo by J. W. Perry)

Figure 90a   Male inflorescence of **corn** (*Zea mays*; *Graminae*), a monocot grass (live, 0.75×). (Photo by J. W. Perry)

Figure 90b   Female inflorescence of **corn**. The protruding stigmas make up the "silk" (live, 0.5×). (Photo by J. W. Perry)

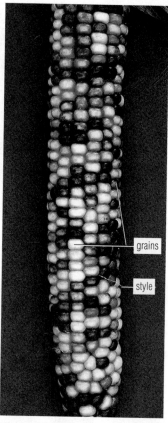

Figure 90c   Corn ear, consisting of a cluster of **fruits** called **grains** attached to the ear axis (cob) (live, 0.5×). (Photo by J. W. Perry)

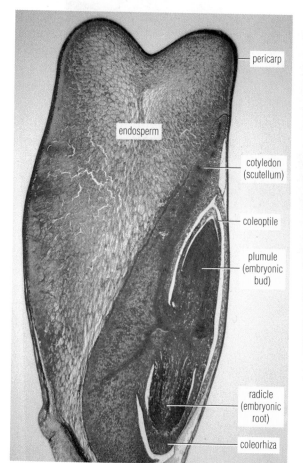

Figure 90d   Corn grain. A **grain** is the one-seeded fruit of a grass. The seed coat is fused to, and virtually impossible to discriminate from, the fruit wall (**pericarp**) (prep. 'e, 7×). (Photo by J. W. Perry)

Figure 90e   Iris (*Iris* sp.; *Iridaceae*) **flower** (live, 0.5×). (Photo by J. W. Perry)

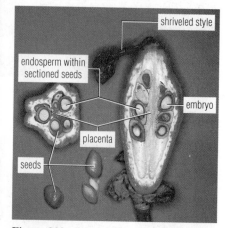

Figure 90f   Iris fruits, called **capsules** (live, c.s. and l.s., 0.5×). (Photo by J. W. Perry)

Figure 91a Inflorescence and immature **fruits** (**berries**) of **palmetto** (*Sabal minor; Arecaceae*). Palmetto leaves are used to thatch roofs (live, 0.1×). (Photo by A. B. Russell)

Figure 91b Mature **fruits** (**berries**) of **palmetto** (live, 0.5×). (Photo by A. B. Russell)

Figure 91c Coconut palm (*Cocos nucifera; Arecaceae*) with **fruits** (**drupes**) (live, 0.15×). (Photo by J. W. Perry)

Figure 91d Mature **coconut palm drupe** with its mesocarp (the husk) intact. Infrequently seen in this form by residents of temperate North America. Typically, the husk has been removed as in Figure 91e (live, l.s., 0.2×). (Photo by J. W. Perry)

Figure 91e Mature **coconut palm drupe** with its mesocarp removed, revealing the stony endocarp (live, l.s., 0.4×). (Photo by J. W. Perry)

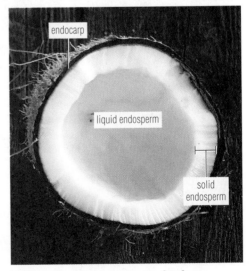

Figure 91f Mature **coconut palm drupe** broken open. The liquid endosperm is sometimes referred to as coconut milk (live, l.s., 0.4×). (Photo by J. W. Perry)

Figure 92a Immature **tomato** (*Lycopersicon esculentum*; *Solanaceae*) **fruits**, **berries**. The U. S. Congress declared that the tomato would be legally considered a vegetable, not a fruit, even though botanically it is a fruit (live, 0.3×). (Photo by J. W. Perry)

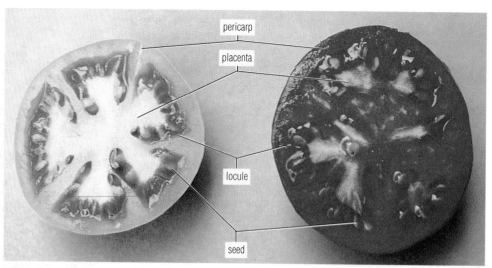

Figure 92b **Immature** (green) and **mature tomato fruits**. During maturation, the placental tissue breaks down, leaving a watery mass (live, c.s., 0.8×). (Photo by J. W. Perry)

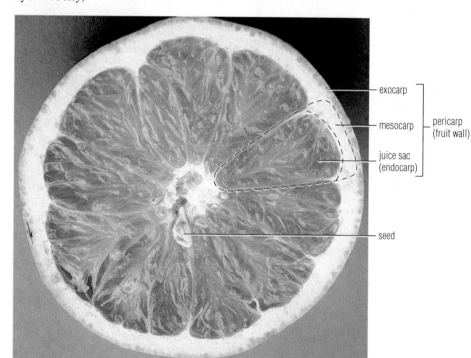

Figure 92c **Grapefruit** (*Citris × paradisi*; *Rutaceae*) **fruit**, a **hesperidium**. The structure of an orange is identical (live, c.s., 0.8×). (Photo by J. W. Perry)

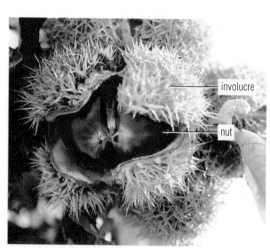

Figure 92d Chinese **chestnut** (*Castanea mollissima*; *Fagaceae*) **fruit**, a **nut**. The spiny involucre (cluster of bracts) is splitting open, exposing the nut (live, 0.75×). (Photo by J. W. Perry)

Figure 93a **Pineapple** (*Ananas comosus; Bromeliaceae*) plant with a young **fruit** (live, 0.15×). (Photo by A. B. Russell)

Figure 93b **Pineapple** fruit, a **multiple fruit** formed from a cluster of closely grouped flowers. Pineapples are seedless fruits (live, 0.4×). (Photo by J. W. Perry)

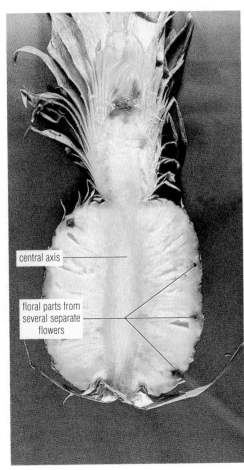

Figure 93c **Pineapple fruit** (live, l.s., 0.4×). (Photo by J. W. Perry)

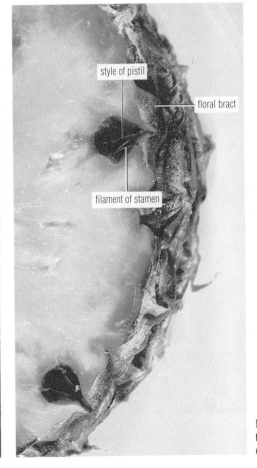

Figure 93d **Pineapple fruit** (live, c.s, 1×). (Photo by J. W. Perry)

Figure 94a **Cancer-root** (*Orobanche uniflora; Orobanchaceae*), a parasite on many kinds of plants (live, 1×). (Photo by J. W. Perry)

Figure 94b **Squaw-root** (*Conopholis americana; Orobanchaceae*), a parasite on the roots of several kinds of trees, including oak (*Quercus*) (live, 0.5×). (Photo by J. W. Perry)

Figure 94c **Beech-drops** (*Epifagus virginiana; Orobanchaceae*), a parasite on beech (*Fagus*) trees (live, 0.5×). (Photo by J. W. Perry)

Figure 94d **Snow plant** (*Sarcodes sanguinea; Pyrolaceae*), so called because it emerges from the snow in the early spring (live, 1×). (Photo by J. W. Perry)

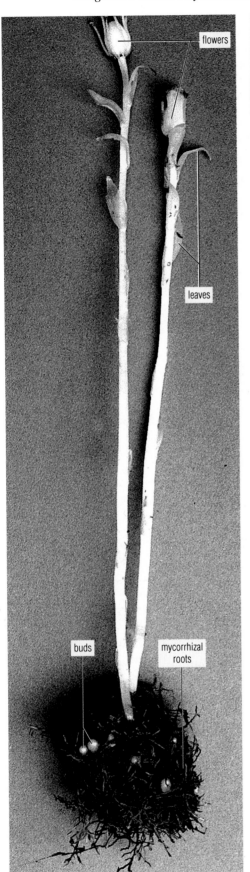

Figure 95a **Indian pipe** (*Monotropa uniflora; Pyrolaceae*), whose mycorrhizal roots link to tree roots to gain nutrition (live, 1.2×). (Photo by J. W. Perry)

Figure 95b **Dwarf mistletoe** (*Arceuthobium* sp.; *Loranthaceae*), an epiparasite on coniferous trees (live, 0.1×). (Photo by J. W. Perry)

Figure 95c **Dwarf mistletoe** (*Arceuthobium* sp.) growing on the branch of a spruce (*Picea* sp.) (live, 1.7×). (Photo by J. D. Davis)

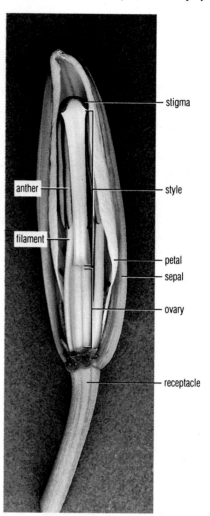

Figure 96a **Lily** flower (Turk's cap lily, *Lilium superbum; Liliaceae*). Sections of floral buds (unexpanded flowers) are used to create slides depicted in Figures 96b through 97b and Figures 98a through 99e (live, 1×). (Photo by J. W. Perry)

Figure 96b **Lily bud** that has been sectioned longitudinally (live, l.s., 1×). (Photo by J. W. Perry)

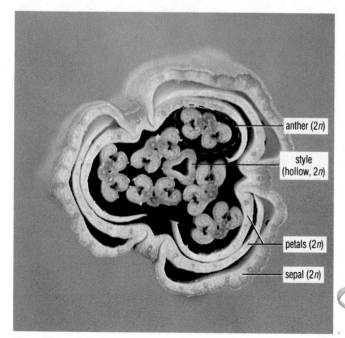

Figure 96c **Lily bud** that has been sectioned transversely (live, c.s., 2.7×). (Photo by J. W. Perry)

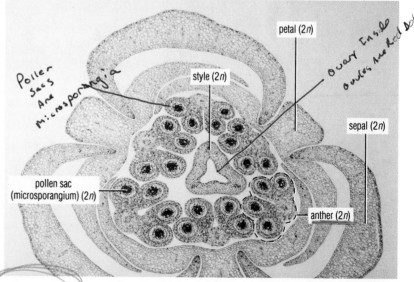

Figure 96d **Lily bud**. Each anther consists of four fused **microsporangia (pollen sacs)** and, at this stage of development, contains diploid microspore mother cells (prep. slide, c.s., 4×). (Photo by J. W. Perry)

Figure 97a Lily anther, containing **pollen grains**. Note that the wall between adjacent pollen sacs has broken down and an opening has developed for the escape of pollen grains (prep. slide, c.s., 26×). (Photo by Biodisc, Inc.)

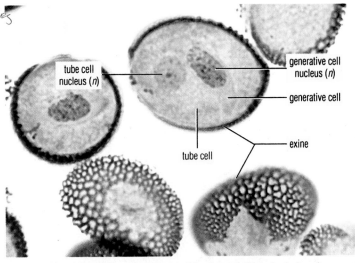

Figure 97b Lily pollen. See Figures 20a through 21g, depicting the events beginning with microspore mother cells and ending with the formation of pollen tetrads (prep. slide, c.s., 275×). (Photo by J. W. Perry)

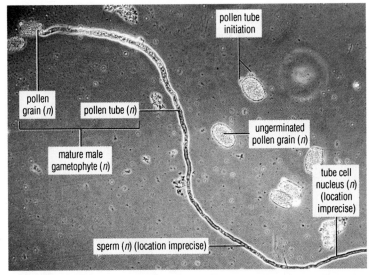

Figure 97c Germinating **pollen grain** of *Impatiens* (live, w.m., 150×). (Photo by J. W. Perry)

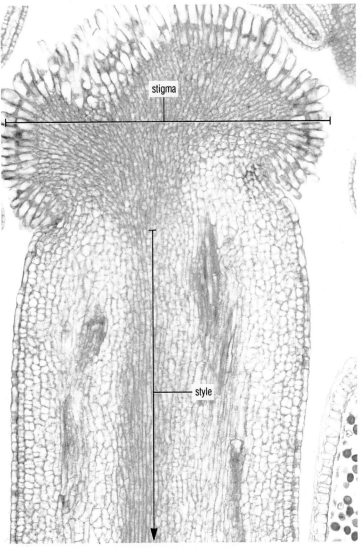

Figure 97d Stigma and style of *Brassica* (prep. slide, l.s., 435×). (Photo by J. W. Perry)

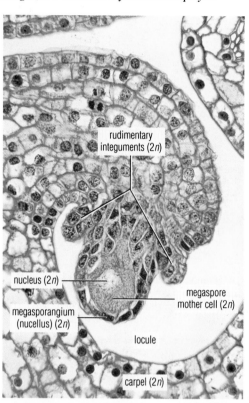

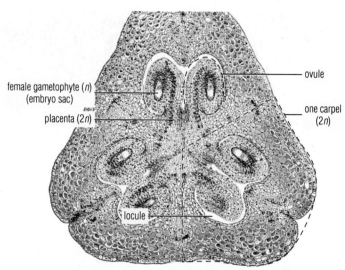

✘ Figure 98a **Lily ovary**. The limit of one carpel is defined by the dashed line, as is one ovule (prep. slide, c.s., 23×). (Photo by J. W. Perry)

Figure 98b **Lily ovule** containing **megaspore mother cell** within the **megasporangium** (nucellus). The ovule in this figure is less developed than those in Figure 98a (prep. slide, c.s., 183×). (Photo by J. W. Perry)

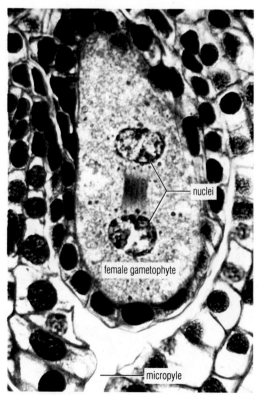

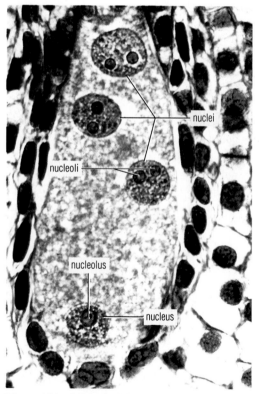

Figure 98c **Lily ovule, two-nucleate stage**. The megaspore mother cell nucleus has just completed the first meiotic division (image shows late telophase), producing a two-nucleate female gametophyte (prep. slide, c.s., 225×). (Photo by W. A. Russin)

Figure 98d **Lily ovule, first four-nucleate stage**. The second meiotic division has produced four haploid (n) nuclei (prep. slide, c.s., 225×). (Photo by W. A. Russin)

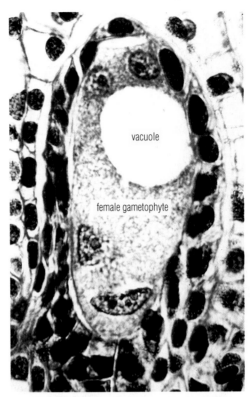

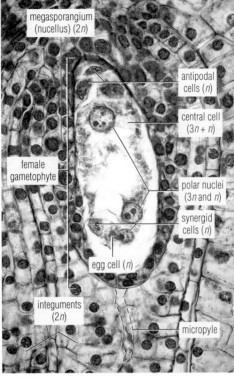

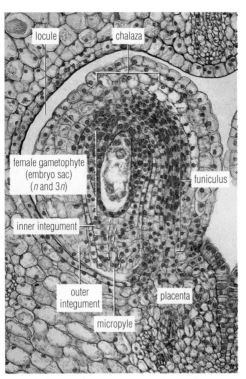

Figure 99a   Lily ovule, second four-nucleate stage. The three nuclei at the chalazal end in Figure 98d have undergone a mitotic division in which all the chromosomes are shared, resulting in two triploid ($3n$) nuclei at the conclusion of that division. The nucleus at the micropylar end underwent mitosis at the same time, producing two haploid (n) nuclei (prep. slide, c.s., 225×). (Photo by W. A. Russin)

Figure 99b   Mature (seven-celled, eight-nucleate) female gametophyte of lily. Another mitotic division has taken place subsequent to the stage shown in Figure 99a to produce this stage. Six of the nuclei have a thin cytoplasm surrounding them (not apparent here). The remaining two polar nuclei are in the central cell. The gametophyte is sometimes referred to as an **embryo sac** (prep. slide, c.s., 225×). (Photo by J. W. Perry)

Figure 99c   Lily ovule containing a **mature female gametophyte**. The location of the micropyle is drawn in with a dashed line (prep. slide, c.s., 85×). (Photo by J. W. Perry)

Figure 99d   "Double fertilization" in lily gametophyte. In actuality, there is only one true fertilization by fusion of one sperm and the egg cell. The other sperm fuses with the two central cell (polar) nuclei (prep. slide, c.s., 215×). (Photo by J. W. Perry)

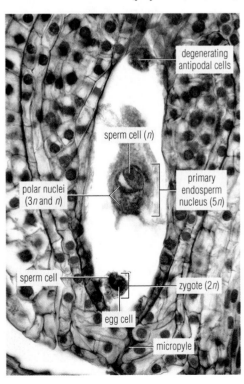

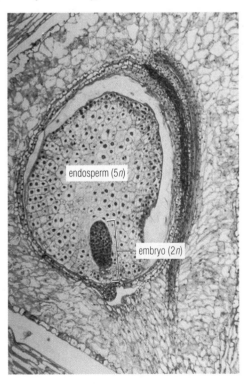

Figure 99e   Lily ovary, with one ovule containing a young **embryo** (prep. slide, c.s., 28×). (Photo by J. W. Perry)

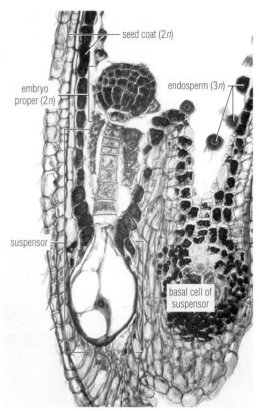

Figure 100a *Capsella* (shepherd's purse, *C. bursa-pastoris; Brassicaceae*) **seed at globular embryo proper stage** (prep. slide, l.s., 230×). (Photo by Biodisc, Inc.)

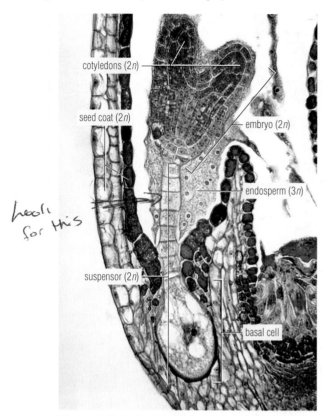

Figure 100b *Capsella* **seed** at stage where **cotyledons** are just developing in the **embryo** (prep. slide, l.s., 215×). (Photo by Biodisc, Inc.)

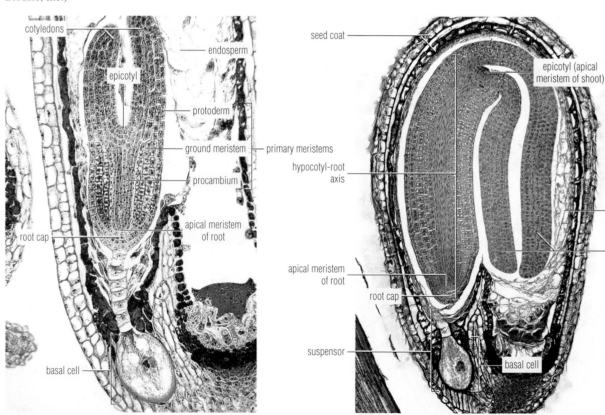

Figure 100c *Capsella* **seed** at stage where **cotyledons** are more fully developed in the **embryo** (prep. slide, l.s., 170×). (Photo by Biodisc, Inc.)

Figure 100d Mature *Capsella* **seed** with **fully formed embryo**. At this stage, the seeds would be shed from the fruit (prep. slide, l.s., 110×). (Photo by Biodisc, Inc.)

Figure 101a *Capsella* with **fruits**, specifically, **capsules** (herbarium specimen, 0.4×). (Photo by J. W. Perry)

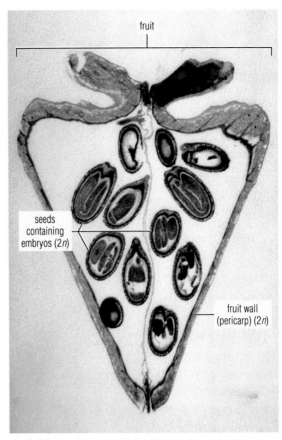

Figure 101b *Capsella* **capsule**, containing **seeds** (prep. slide, 50×). (Photo by J. W. Perry)

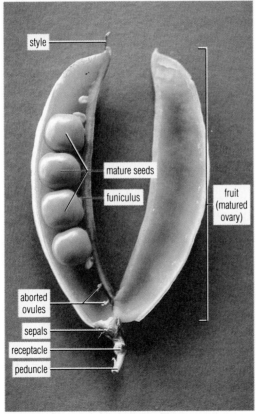

Figure 101c **Fruit** (a **pod**) containing seeds of **garden pea** (*Pisum sativum*) (live, 0.9×). (Photo by J. W. Perry)

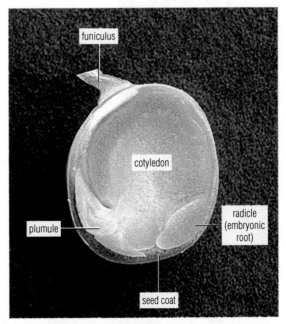

Figure 101d **Garden pea seed** (live, sec., 3.5×). (Photo by J. W. Perry)

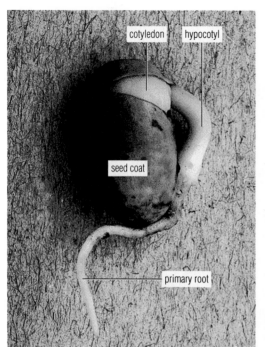

Figure 102a **Germinating bean seed** (*Phaseolus vulgaris*), one day after sowing seed (live, 3×). (Photo by J. W. Perry)

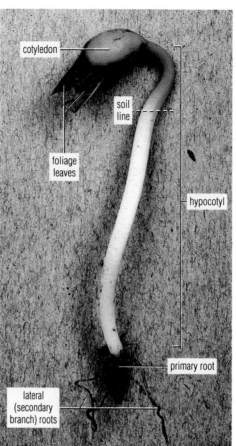

Figure 102b **Bean seedling**, three or four days after sowing seed. The seed coat has been shed (live, 1×). (Photo by J. W. Perry)

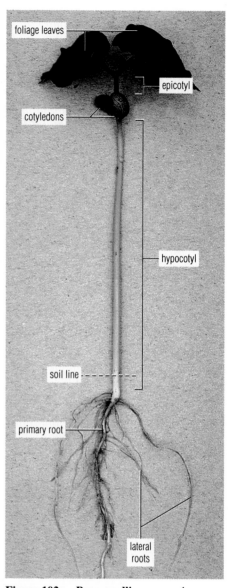

Figure 102c **Bean seedling**, seven days after sowing seed (live, 0.75×). (Photo by J. W. Perry)

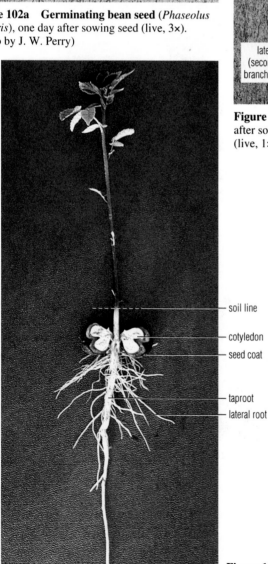

Figure 102d **Black walnut** (*Juglans nigra*) **seedling** (live, 0.5×). (Photo by J. W. Perry)

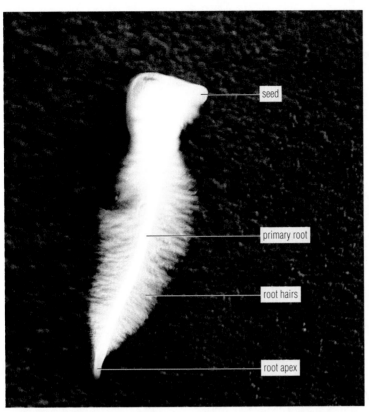

Figure 103a **Root hairs** on primary root of germinating **radish** (*Raphanus sativus*) seed (live, 6×). (Photo by J. W. Perry)

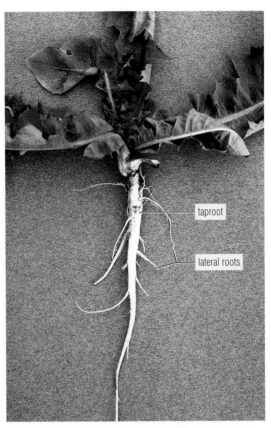

Figure 103b **Taproot system** of dandelion (*Taraxacum officinale*). See also black walnut seedling in Figure 102d (live, 0.5×). (Photo by J. W. Perry)

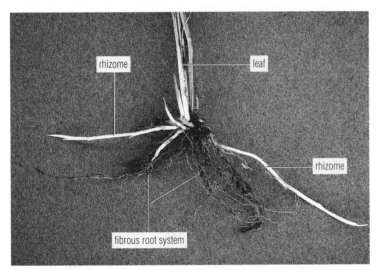

Figure 103c **Fibrous root system** of quack grass (*Agropyron repens*). Quack grass has **rhizomes** (underground stems) that allow it to spread aggressively (live, 0.5×). (Photo by J. W. Perry)

Figure 103d **Prop roots** of corn (*Zea mays*). These are **adventitious roots**, arising from the stem (live, 0.5×). (Photo by J. W. Perry)

Figure 104a   Adventitious roots arising from stem cutting of geranium (*Pelargonium hortorum*) (live, 0.9×). (Photo by J. W. Perry)

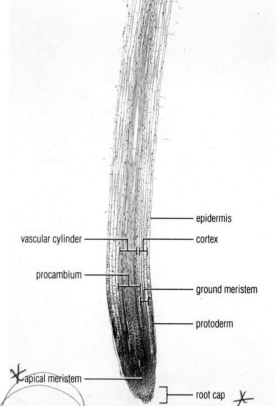

epidermis

vascular cylinder — cortex

procambium

ground meristem

protoderm

apical meristem

root cap

Final

Figure 104b   Onion (*Allium*) **root** (prep. slide, l.s., 25×). (Photo by J. W. Perry)

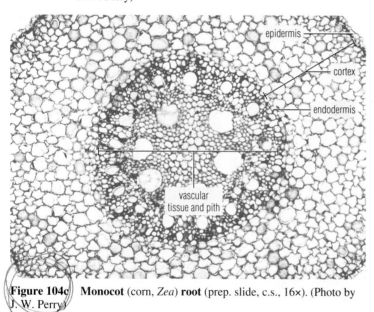

epidermis

cortex

endodermis

vascular tissue and pith

Figure 104c   Monocot (corn, *Zea*) **root** (prep. slide, c.s., 16×). (Photo by J. W. Perry)

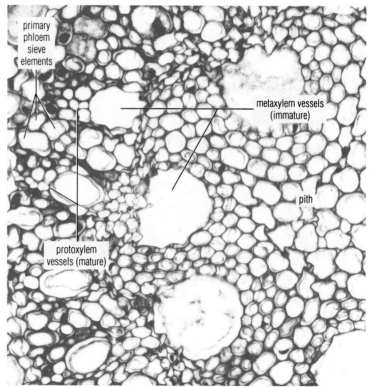

primary phloem sieve elements

metaxylem vessels (immature)

pith

protoxylem vessels (mature)

Figure 104d   Portion of **vascular cylinder** of a **monocot** (corn, *Zea*) **root** (prep. slide, c.s., 140×). (Photo by J. W. Perry)

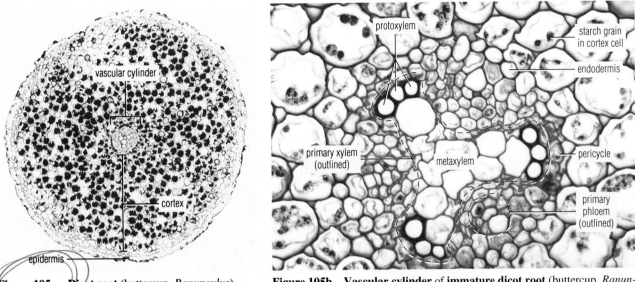

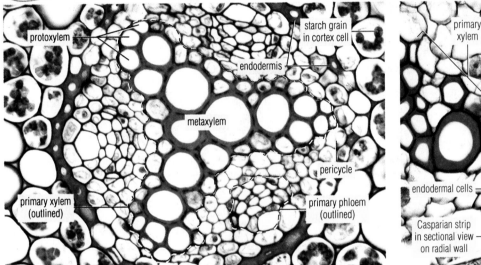

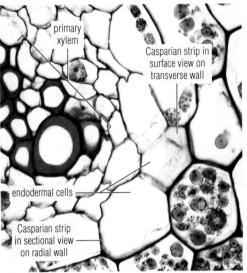

Figure 105a **Dicot root** (buttercup, *Ranunculus*) (prep. slide, c.s., 37×). (Photo by J. W. Perry)

Figure 105b **Vascular cylinder** of **immature dicot root** (buttercup, *Ranunculus*) (prep. slide, c.s., 390×). (Photo by J. W. Perry)

Figure 105c **Vascular cylinder** of **mature dicot root** (buttercup, *Ranunculus*) (prep. slide, c.s., 390×). (Photo by J. W. Perry)

Figure 105d **Casparian strip** in **root endodermal cells** (prep. slide, c.s., 370×). (Photo by J. W. Perry)

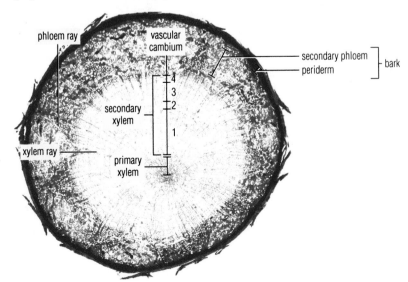

Figure 105e **Woody dicot root** (basswood, *Tilia*) (prep. slide, c.s., 20×). (Photo by J. W. Perry)

Figure 106a **Mycorrhizal roots** of an unidentified conifer (live, 5×).

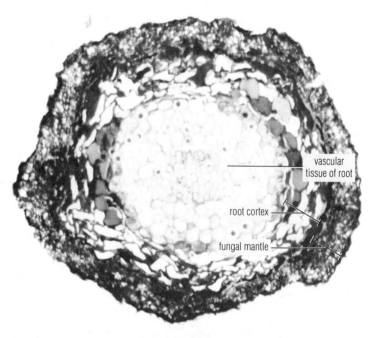

Figure 106b **Ectomycorrhizal root** (*Pinus*) (prep. slide, c.s., 30×). (Photo by J. W. Perry)

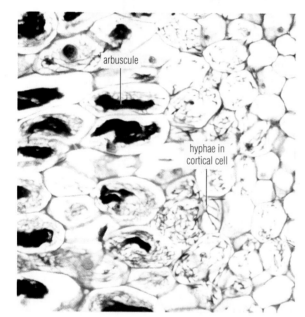

Figure 106c **Endomycorrhizal root** (*Corallorhiza*) (prep. slide, c.s., 110×). (Photo by J. W. Perry)

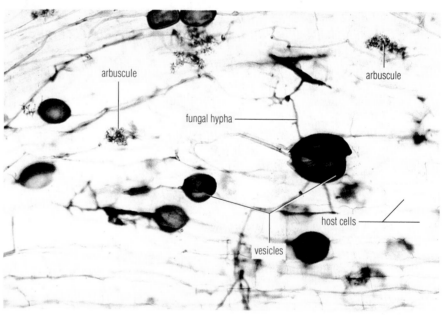

Figure 106d **Vesicles** and **arbuscules** of an **endomycorrhiza**. The fungus in this root is *Glomus* (prep. slide, squash, 150×). (Photo by J. W. Perry)

Figure 107a   Lateral (branch = secondary) roots arising from primary roots of water lettuce (*Pistia stratiotes*) (live, 1×). (Photo by J. W. Perry)

lateral (branch) roots

primary roots

Figure 107c   Branch root that has broken through the primary root's surface (willow, *Salix*) (prep. slide, c.s., 51×). (Photo by J. W. Perry)

Figure 107b   Early stage of branch root formation, arising from pericycle of a willow (*Salix*) root (prep. slide, c.s., 56×). (Photo by J. W. Perry)

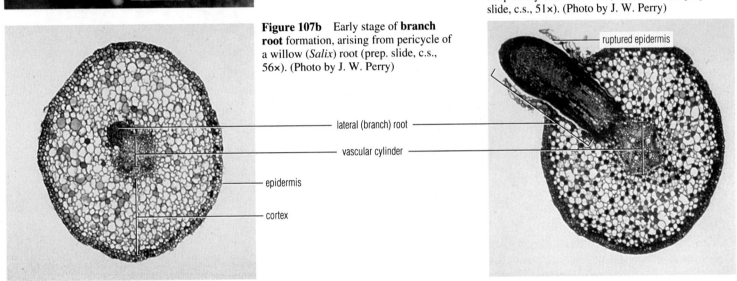

ruptured epidermis

lateral (branch) root

vascular cylinder

epidermis

cortex

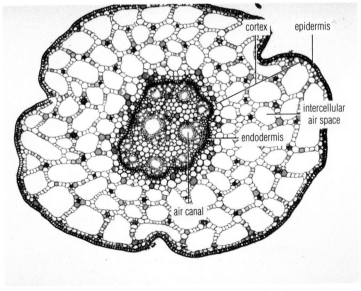

velamen

endodermis

cortex

vascular cylinder

cortex

epidermis

intercellular air space

endodermis

air canal

Figure 107d   Aerial root with its multilayered, thick-walled epidermis called the **velamen**, which serves as a storage and protective tissue (prep. slide, c.s., 36×). (Photo by J. W. Perry)

Figure 107e   Water root, with large intercellular gas chambers necessary for the accumulation and diffusion of oxygen required for aerobic cellular respiration of tissues with no direct access to the air (prep. slide, c.s., 40×). (Photo by J. W. Perry)

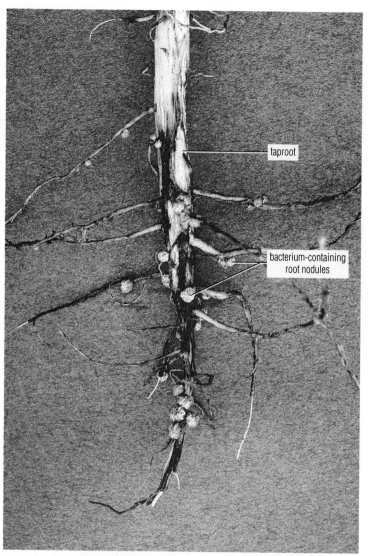

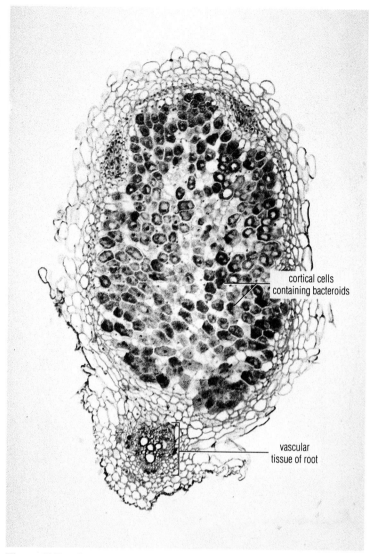

Figure 108a Root system of soybean (*Glycine*), with **root nodules** containing nitrogen-fixing bacteria of the genus *Rhizobium*. Circled nodule is the subject of Figure 108b (live, 1.2×). (Photo by J. W. Perry)

Figure 108b **Root nodule** of a legume such as soybean or pea plants (prep. slide, c.s., 55×). (Photo by J. W. Perry)

Figure 109a *Coleus* **shoot tip** with leaves numbered according to age: 1 is youngest, 4 oldest (live, 1×). (Photo by J. W. Perry)

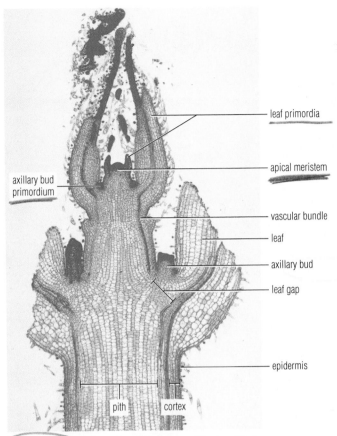

leaf primordia

apical meristem

axillary bud primordium

vascular bundle

leaf

axillary bud

leaf gap

epidermis

pith   cortex

Figure 109b *Coleus* **shoot tip** (prep. slide, l.s., 27×). (Photo by J. W. Perry)

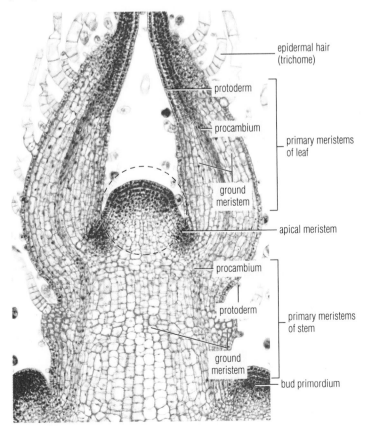

epidermal hair (trichome)

protoderm

procambium

primary meristems of leaf

ground meristem

apical meristem

procambium

protoderm

primary meristems of stem

ground meristem

bud primordium

Figure 109c *Coleus* **shoot apex** before leaf primordium formation by apical meristem (prep. slide, l.s., 110×). (Photo by J. W. Perry)

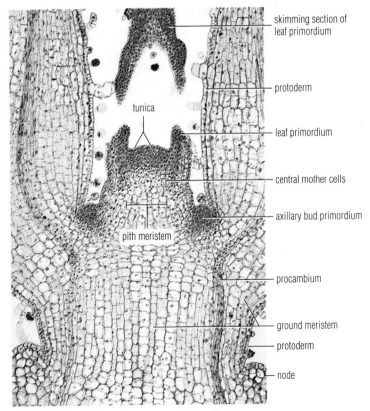

skimming section of leaf primordium

protoderm

tunica

leaf primordium

central mother cells

axillary bud primordium

pith meristem

procambium

ground meristem

protoderm

node

Figure 109d *Coleus* **shoot apex** with **leaf primordia** formed by apical meristem. Labels in this figure call out more detail than those in Figure 109c (prep. slide, l.s., 245×). (Photo by J. W. Perry)

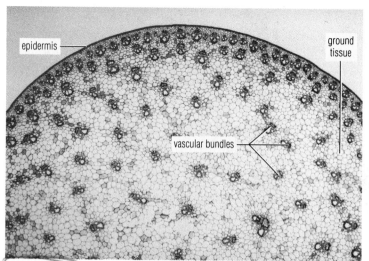

Figure 110a **Monocot stem** (corn, *Zea*) (prep. slide, c.s., 15×). (Photo by J. W. Perry)

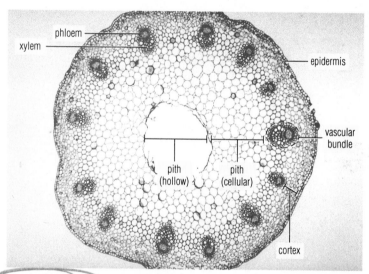

Figure 110c **Herbaceous dicot stem** with a pith that is partially hollow due to cellular tearing during stem elongation (buttercup, *Ranunculus*) (prep. slide, c.s., 30×). (Photo by J. W. Perry)

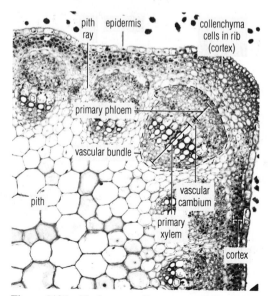

Figure 110d **Herbaceous dicot stem** (alfalfa, *Medicago*) (prep. slide, c.s., 100×). (Photo by J. W. Perry)

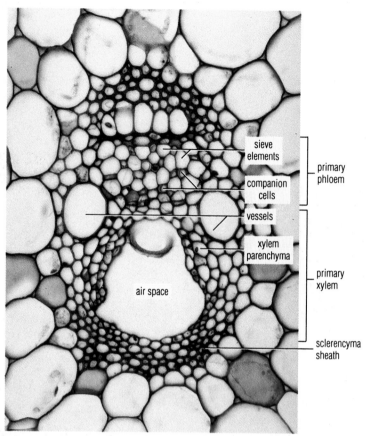

Figure 110b **Vascular bundle** of **monocot stem** (corn, *Zea*) (prep. slide, c.s., 300×). (Photo by J. W. Perry)

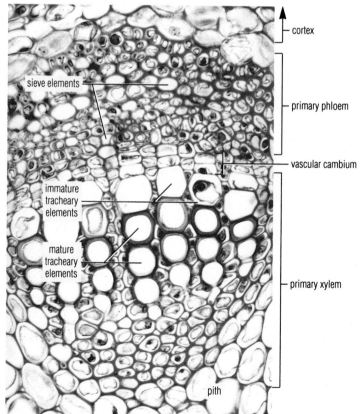

Figure 110e **Vascular bundle** of **herbaceous dicot stem** (alfalfa, *Medicago*) (prep. slide, c.s., 410×). (Photo by J. W. Perry)

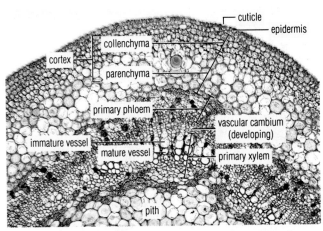

Figure 111a   Woody dicot stem (basswood, *Tilia*) during first month's seedling growth. This stage is typically labeled "young stem" (prep. slide, c.s., 20×). (Photo by J. W. Perry)

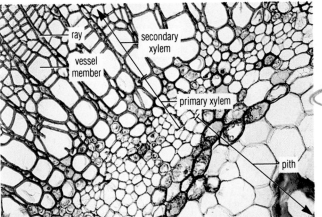

Figure 111c   Pith, primary xylem, and **secondary xylem** of **woody dicot stem** (basswood, *Tilia*) shown in Figure 111b (prep. slide, c.s., 80×). (Photo by J. W. Perry)

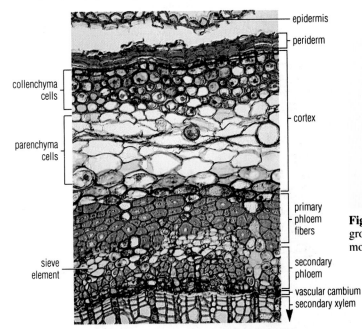

Figure 111d   Secondary xylem, vascular cambium, phloem, cortex, young periderm, and **epidermis** of **woody dicot stem** (basswood, *Tilia*) shown in Figure 111b (prep. slide, c.s., 80×). (Photo by J. W. Perry)

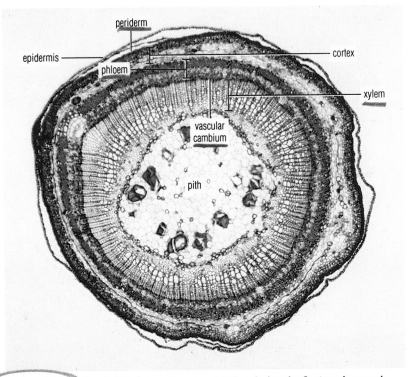

Figure 111b   Woody dicot stem (basswood, *Tilia*) during the first year's growth. The seedling would be about three months old at this stage (prep. slide, c.s., 22×). (Photo by J. W. Perry)

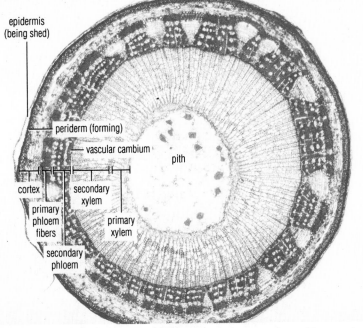

Figure 111e   Woody dicot stem (basswood, *Tilia*) at completion of first year's growth (typically labeled "one-year stem"). The seedling would be about seven months old at this stage (prep. slide, c.s., 15×). (Photo by J. W. Perry)

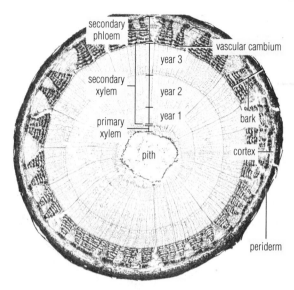

Figure 112a **Woody dicot stem** (basswood, *Tilia*) at completion of third year's growth (prep. slide, c.s., 10×). (Photo by J. W. Perry)

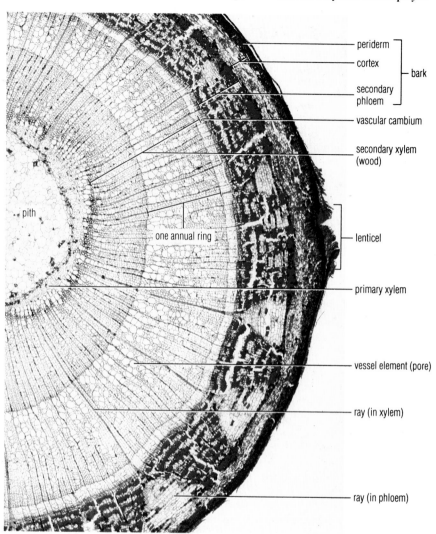

Figure 112b **Woody dicot stem** (basswood, *Tilia*) that is three years old. This stem was collected early in the growing season and, as a consequence, the third annual ring is much narrower than the first two, because growth for the third year had not been completed (prep. slide, c.s., 20×). (Photo by J. W. Perry)

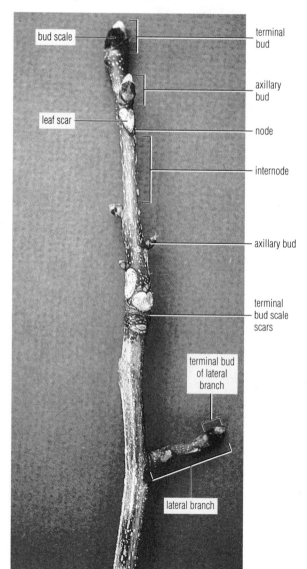

Figure 112c **Woody dicot branch** (hickory, *Carya*) (live, 0.8×). (Photo by J. W. Perry)

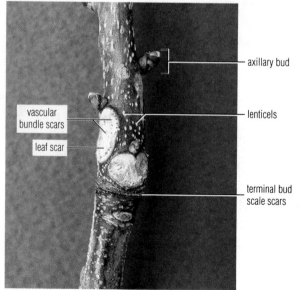

Figure 112d **Woody dicot branch** (hickory, *Carya*) (live, 1×). (Photo by J. W. Perry)

Figure 113a **Bark** of a relatively young black cherry tree (*Prunus serotina*) with prominent **lenticels** (live, 0.4×). (Photo by J. W. Perry)

Figure 113b **Bark** of an older black cherry tree (*Prunus serotina*) with **scaly bark** that has replaced the smooth bark shown in Figure 113a (live, 0.4×). (Photo by J. W. Perry)

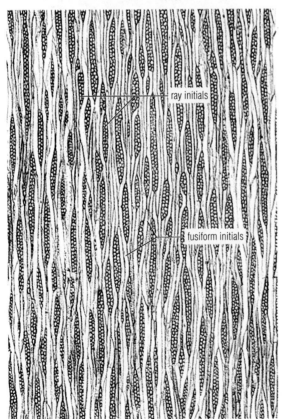

Figure 113c **Vascular cambium** of apple (*Malus*). This is a nonstoried cambium (compare with Figure 113d) (prep. slide, t.s., 50×). (Photo by J. W. Perry)

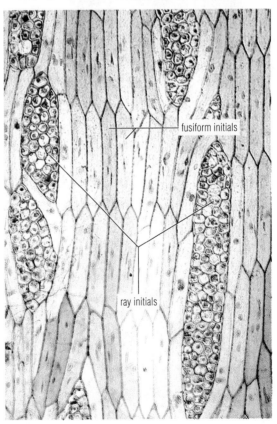

Figure 113d **Vascular cambium** of locust (*Robinia*). This is a storied cambium (compare with Figure 113c) (prep. slide, t.s., 160×). (Photo by J. W. Perry)

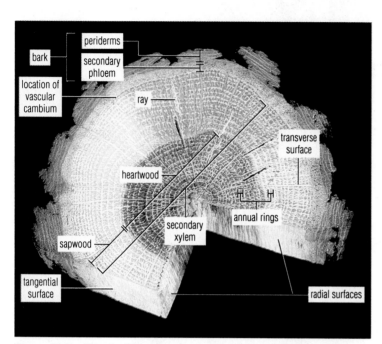

Figure 114a   Woody dicot stem (bur oak, *Quercus macrocarpa*) that is about 27 years old. Within both the heartwood and the sapwood, the lighter cells are the early wood and the darker cells the late wood. This stem was cut to reveal transverse, radial, and tangential surfaces (live, c.s., 1×). (Photo by J. W. Perry)

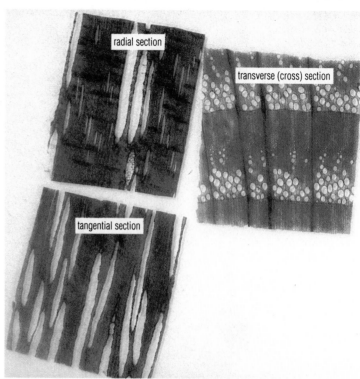

Figure 114b   Oak wood (*Quercus*), **radial**, **transverse** (cross), **and tangential sections** (prep. slide, 6×). (Photo by J. W. Perry)

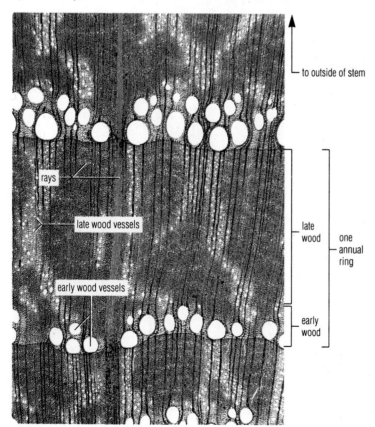

Figure 114c   Oak wood (*Quercus*), **transverse section** showing several growth increments ("annual rings"). This is a **ring-porous wood** (prep. slide, c.s., 27×). (Photo by J. W. Perry)

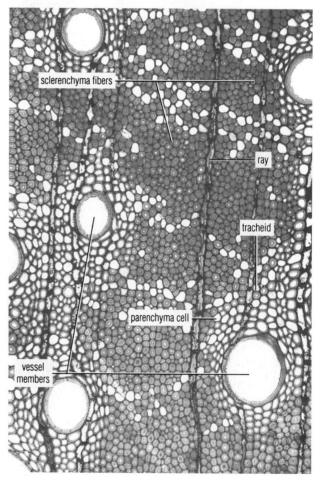

Figure 114d Portion of a single growth increment of **oak wood** (*Quercus*), as seen in **transverse section** (prep. slide, c.s., 90×). (Photo by J. W. Perry)

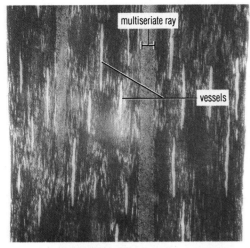

Figure 115a   Oak wood (*Quercus*), **tangential section** (prep. slide, t.s., 4×). (Photo by J. W. Perry)

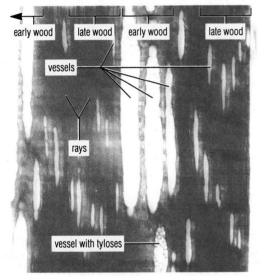

Figure 115b   Oak wood (*Quercus*), **radial section**. This section consists of two growth increments (prep. slide, r.s., 5×). (Photo by J. W. Perry)

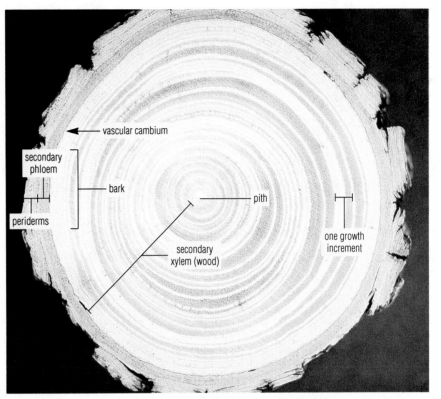

Figure 115c   Ash stem (*Fraxinus*), a woody dicot. Compare this with oak (Figure 114a), noting the absence of prominent heartwood in ash (prep. slide, c.s., 1.2×). (Photo by J. W. Perry)

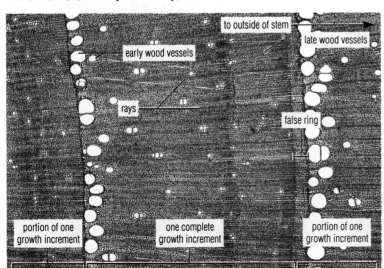

Figure 115d   Ash wood (*Fraxinus*), a **ring-porous wood**. Note the differences in wood density in the center annual ring, a consequence of environmental factors such as water availability at different times of the growing season affecting wood production and growth (prep. slide, c.s., 24×). (Photo by J. W. Perry)

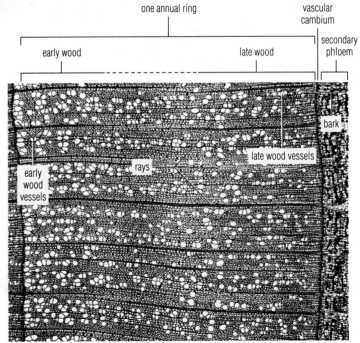

Figure 115e   Basswood wood (*Tilia*), a **diffuse-porous wood**. The transition from early wood to late wood is much less obvious than in oak (prep. slide, c.s., 24×). (Photo by J. W. Perry)

Figure 116a   Woody conifer stem (red pine, *Pinus resinosa*) that is seven years old. This branch is the same diameter as the 27-year-old bur oak stem in Figure 114a and 25+-year-old ash branch in Figure 115d, demonstrating the rapid growth rate of red pine compared to bur oak and ash (live, c.s., 0.9×). (Photo by J. W. Perry)

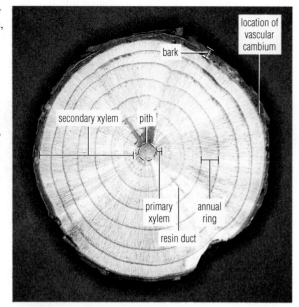

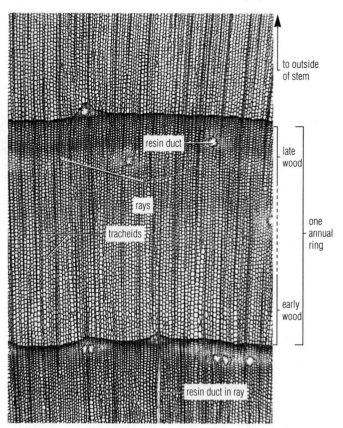

Figure 116b   Pine wood (*Pinus*), **transverse** (cross), **radial, and tangential sections** (prep. slide, t.s., 5×). (Photo by J. W. Perry)

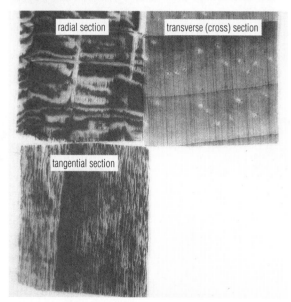

Figure 116c   Pine wood (*Pinus*), **transverse section**. Lacking vessels, this is **nonporous wood** (prep. slide, c.s., 25×). (Photo by J. W. Perry)

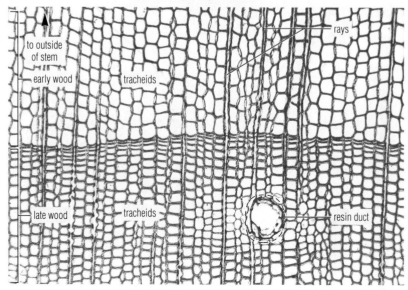

Figure 116d   Pine wood (*Pinus*), **transverse section** with portions of two growth increments (prep. slide, c.s., 80×). (Photo by J. W. Perry)

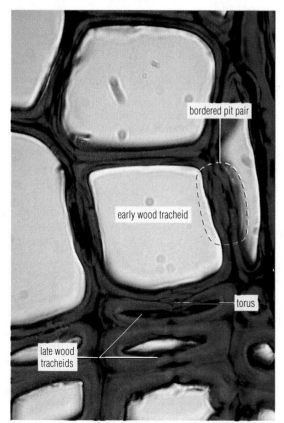

Figure 116e   Pine wood (*Pinus*) showing **tracheids** at junction of early and late wood with **bordered pit pairs** (prep. slide, c.s., 960×). (Photo by J. W. Perry)

to outside of stem ⟶

early wood   late wood   early wood

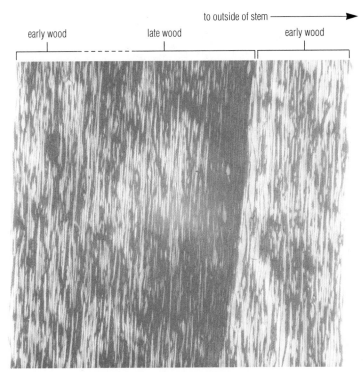

Figure 117a   **Pine wood** (*Pinus*), **tangential section** (prep. slide, t.s., 6×). (Photo by J. W. Perry)

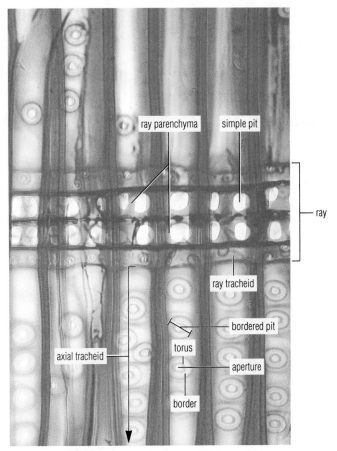

Figure 117c   **Pine wood** (*Pinus*), **radial section** showing detail of **tracheids** and **ray** (prep. slide, r.s., 370×). (Photo by J. W. Perry)

late wood   early wood   to outside of stem ⟶

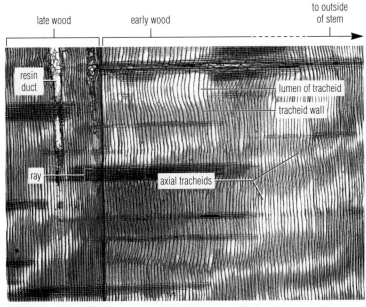

Figure 117b   **Pine wood** (*Pinus*), **radial section**. The light-dark banding is due to the tracheids being cut through their lumens (light) or surface of cell wall (dark) (prep. slide, r.s., 23×). (Photo by J. W. Perry)

Figure 118a   **Simple leaves** of a pear plant (*Pyrus*) (live, 1×). (Photo by J. W. Perry)

Figure 118b   **Palmately compound leaves** of horse chestnut (*Aesculus hippocastanum*) (live, 0.3×). (Photo by J. W. Perry)

Figure 118c   **Pinnately compound leaves** of mountain ash (*Sorbus aucuparia*) (live, 1×). (Photo by J. W. Perry)

Figure 118d   **Sessile leaf** of *Moricandia* (live, 1.3×). (Photo by J. W. Perry)

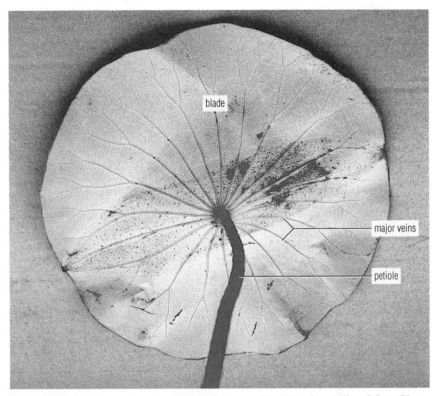

Figure 119a   Sessile sheathing leaf typical of grasses. Note the **parallel venation** characteristic of many monocots (corn, *Zea mays*) (live, 1×). (Photo by J. W. Perry)

Figure 119b   Lower surface of **peltate leaf** of water lily (*Nymphaea*) (live, 0.7×). (Photo by J. W. Perry)

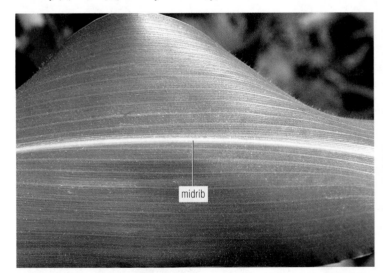

Figure 119c   Monocot leaf (corn, *Zea mays*) with characteristic **parallel venation** (live, 1×). (Photo by J. W. Perry)

Figure 119d   Dicot leaf (fig, *Ficus* sp.) with characteristic **netted venation** (live, 0.4×). (Photo by J. W. Perry)

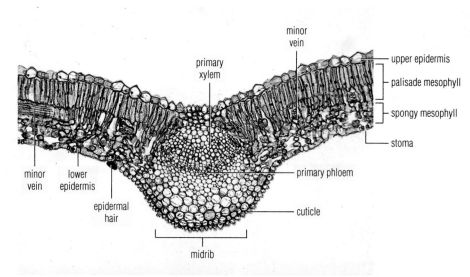

Figure 120a **Dicot leaf** (lilac, *Syringa*), a **mesomorphic leaf** (prep. slide, c.s., 24×). (Photo by J. W. Perry)

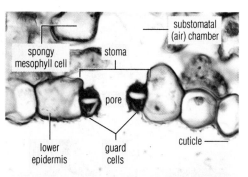

Figure 120b **Stoma** in lower epidermis of **dicot leaf** (lilac, *Syringa*) (prep. slide, c.s., 600×). (Photo by J. W. Perry)

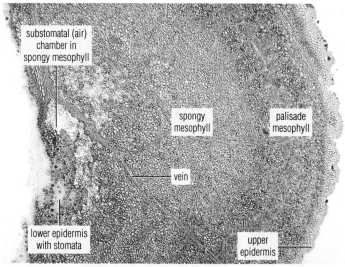

Figure 120c **Oblique section** running from upper epidermis to lower epidermis of a **dicot leaf** (lilac, *Syringa*). Slide labels typically say "**paradermal section**," which is inaccurate because a paradermal section is one that runs parallel to the surface (prep. slide, 24×). (Photo by J. W. Perry)

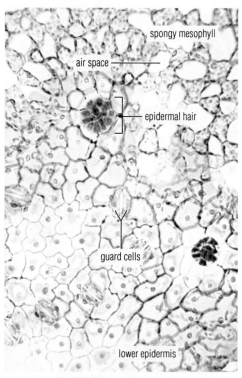

Figure 120d Portion of section from Figure 120c (prep. slide, oblique section, 116×). (Photo by J. W. Perry)

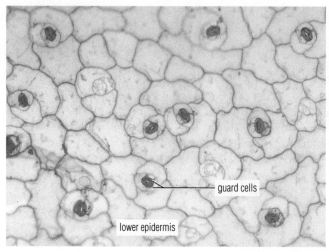

Figure 120e **Dicot leaf epidermis** (prep. slide, w.m., 93×). (Photo by J. W. Perry)

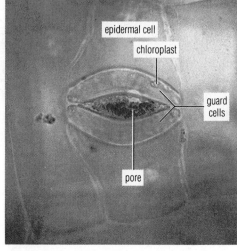

Figure 120f **Stoma** of a **dicot leaf** (*Zebrina*) (live, w.m., 230×). (Photo by J. W. Perry)

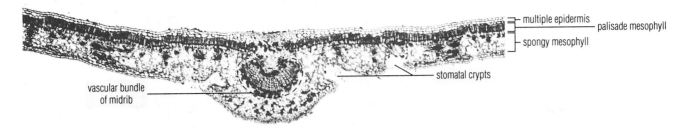

Figure 121a   Xeromorphic leaf (oleander, *Nerium*) (prep. slide, c.s., 19×). (Photo by J. W. Perry)

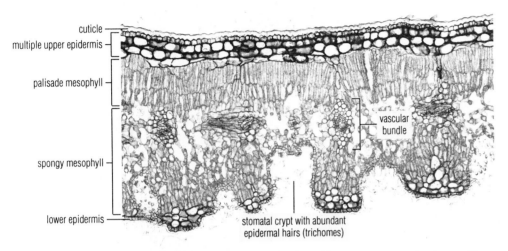

Figure 121b   Xeromorphic leaf, adapted for water conservation in an arid environment (oleander, *Nerium*), showing lower epidermis **stomatal crypts** (prep. slide, c.s., 87×). (Photo by J. W. Perry)

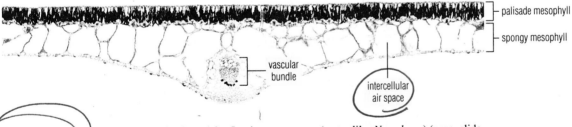

Figure 121c   Hydromorphic leaf, adapted for floating atop water (water lily, *Nymphaea*) (prep. slide, c.s., 31×). (Photo by J. W. Perry)

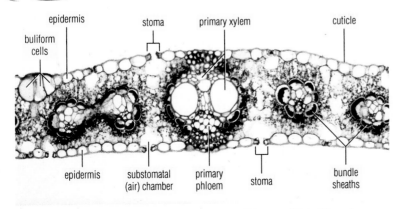

Figure 121d   Monocot leaf (corn, *Zea*) (prep. slide, c.s., 97×). (Photo by J. W. Perry)

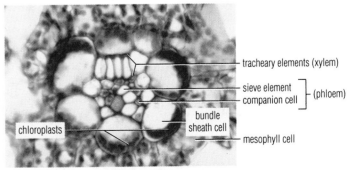

Figure 121e   Vascular bundle showing **Kranz anatomy**, typical of C$_4$ grasses (corn, *Zea mays*) (prep. slide, c.s., 232×). (Photo by J. W. Perry)

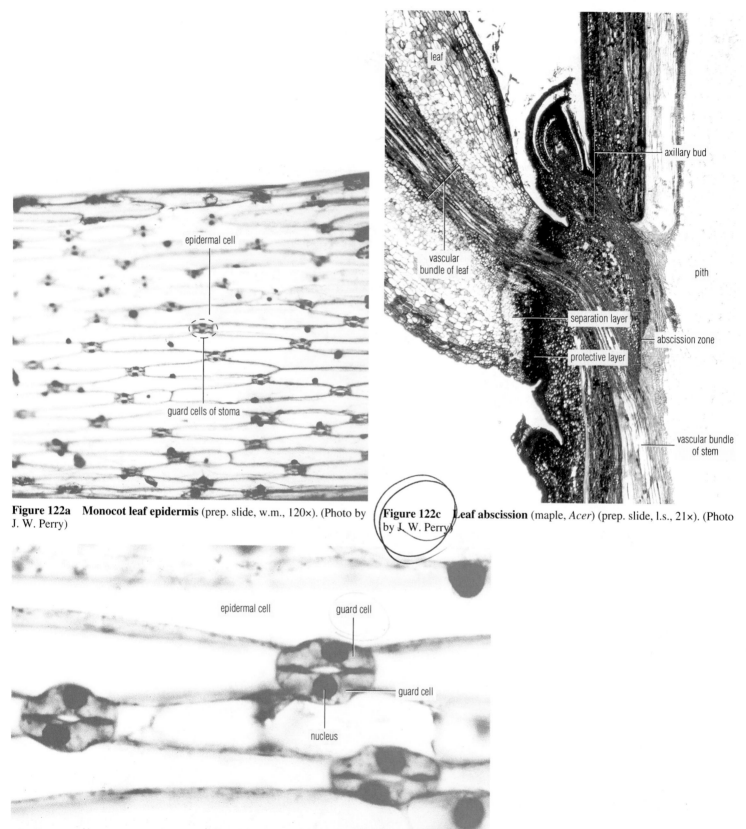

Figure 122a **Monocot leaf epidermis** (prep. slide, w.m., 120×). (Photo by J. W. Perry)

Figure 122c **Leaf abscission** (maple, *Acer*) (prep. slide, l.s., 21×). (Photo by J. W. Perry)

Figure 122b **Stomata** of a grass leaf (corn, *Zea*) (prep. slide, w.m., 650×). (Photo by J. W. Perry)

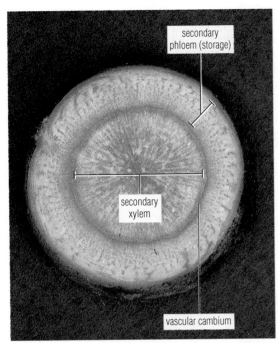

Figure 123a   Carrot taproot (*Daucus carota*), a storage root. This image is from a large taproot, about 5 cm in diameter. You can see from its anatomy why it would not be preferred for eating—too much wood (live, c.s., 0.9×). (Photo by J. W. Perry)

Figure 123b   Pneumatophores of bald cypress (*Taxodium distichum*), adapted for obtaining atmospheric oxygen (live, 0.01×). (Photo by J. W. Perry)

Figure 123c   Aerial roots, adapted for clinging to trees upon which these epiphytes grow. Refer to Figure 107d for the anatomy of an aerial orchid root (live, 0.2×). (Photo by J. W. Perry)

Figure 123d   Buttress roots, an adaptation for stabilizing tall trees growing in shallow soils, such as this Belizian kapok tree (*Ceiba pentandra*) (live, 0.01×). (Photo by J. W. Perry)

Figure 124a **Rhizome** of iris (*Iris*), adapted for vegetative reproduction (live, 0.3×). (Photo by J. W. Perry)

Figure 124b **Tubers** of potato (*Solanum tuberosum*), adapted for vegetative reproduction and accumulation of starch (live, 0.3×). (Photo by J. W. Perry)

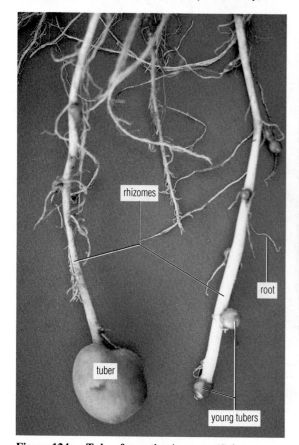

Figure 124c **Tuber formation** in potato (*Solanum tuberosum*) (live, 0.5×). (Photo by J. W. Perry)

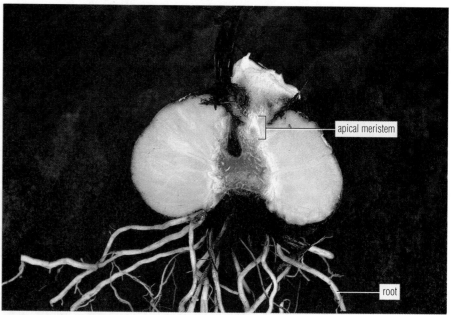

Figure 124d Gladiolus (*Gladiolus*) **corm**, an underground stem that accumulates food reserves, typically in the form of starch (live, l.s., 1×). (Photo by J. W. Perry)

Figure 125a Storage stem of kohlrabi (*Brassica olerace*), adapted for storage of carbohydrates (live, 0.5×). (Photo by J. W. Perry)

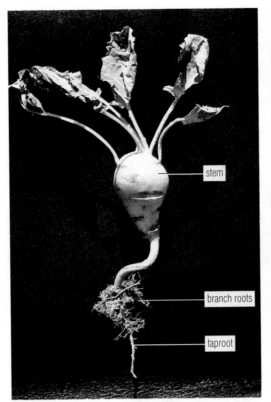

stem

branch roots

taproot

Figure 125b Thorns of hawthorn (*Crataegus*), adapted for protection against large herbivores. These are modified branches, arising from axillary buds (live, 0.4×). (Photo by J. W. Perry)

thorns

Figure 125c Cladophylls of *Cryptocerus*, adapted as the plant's primary photosynthetic organs (live, 0.2×). (Photo by J. W. Perry)

Figure 125d Cladophylls of asparagus (*Asparagus*), adapted for photosynthesis (live, 0.3×). (Photo by J. W. Perry)

tendrils

Figure 125e Grape (*Vitis*) **tendrils** are stems modified to wrap around nearby objects and support the plant (live, 0.3×). (Photo by J. W. Perry)

Figure 126a **Tendrils** of garden peas (*Pisum sativum*) are the terminal portion of pinnately compound leaves. Like the tendrils of grape (Figure 125e), they function for support (live, 1×). (Photo by J. W. Perry)

Figure 126c Cactus (*Echinocactus grusonii*) **spines**, leaves that have been modified to ward off animals that might be seeking water stores within the cactus (live, 0.5×). (Photo by J. W. Perry)

Figure 126b **Leaves** comprising the **bulb** of an onion (*Allium cepa*) are modified for storage of food reserves used when the bulb begins new growth (live, 0.8×). (Photo by J. W. Perry)

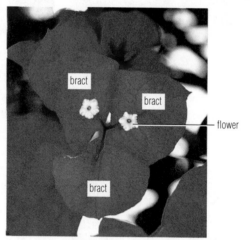

Figure 126d **Bracts** that supplement the otherwise inconspicuous flowers of *Bougainvillea* (live, 0.5×). (Photo by J. W. Perry)

Figure 126e **Bracts** of poinsettia (*Euphorbia pulcherrima*). These bracts turn from green to red under long night (short day) conditions. People lacking knowledge in botany would probably call them "flowers" (live, 0.5×). (Photo by J. W. Perry)

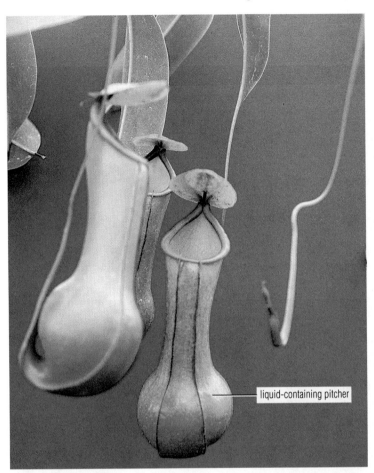

Figure 127b   Insect-trapping leaves of a sundew (*Drosera*). The leaves produce sticky, glandular hairs to which small insects adhere (live, 1×). (Photo by J. W. Perry)

Figure 127a   Pitcher leaves of plant (*Nepenthes*) adapted for insect trapping (live, 0.8×). (Photo by J. W. Perry)

Figure 127c   Insect traps made of modified leaves in Venus's-flytrap (*Dionaea muscipula*). Sensitive hairs within the trap respond to touch, resulting in closure of the trap (live, 1×). (Photo by J. W. Perry)

Figure 127d   Leaf with drip tip, an adaptation of rain-forest plant leaves that allows them to shed water quickly. Quick drying permits the leaf to begin photosynthesizing sooner than it would if it remained wet; it also prevents growth of algae, fungi, and lichens that could cover the leaf surface (live, 1×). (Photo by J. W. Perry)

Figure 128a **Plantlets** produced by the leaves of *Kalanchoe* (live, 0.8×). (Photo by J. W. Perry)

Figure 128b **Plantlets** produced by the leaves of the mother fern (*Asplenium bulbiferum*) (live, 0.8×). (Photo by J. W. Perry)

Figure 128c **Stolon** of strawberry (*Fragaria*) (live, 0.4×). (Photo by J. W. Perry)

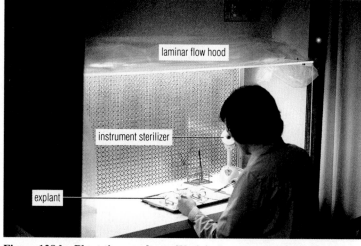

Figure 128d **Plant tissue culture**. Work is done in a sterile laminar flow transfer hood to prevent bacterial contamination of the explants (living tissue removed and placed in artificial culture medium). (Photo by J. W. Perry)

Figure 128e **Tissue culture explants** are grown on a sterile agar medium that contains necessary hormones and nutrients. (Photo by J. W. Perry)

Figure 128f **Tissue culture explants** ready to be sent to a nursery, where they will be potted and grown in a soil medium. These are genetically identical plants (live, 1×). (Photo by J. W. Perry)

Note: See Figures 124a–d and 126b for other examples of vegetative propagation.

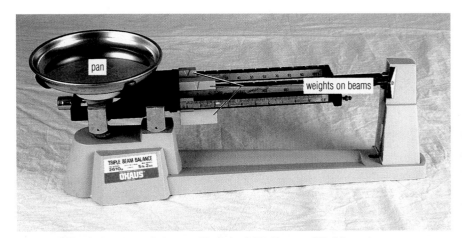

Figure 129a   Triple beam balance. (Photo by J. W. Perry and D. Morton)

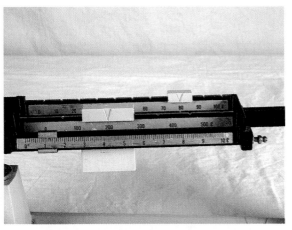

Figure 129b   Reading the scale on a triple beam balance. The correct reading is 281.1 g. (Photo by J. W. Perry and D. Morton)

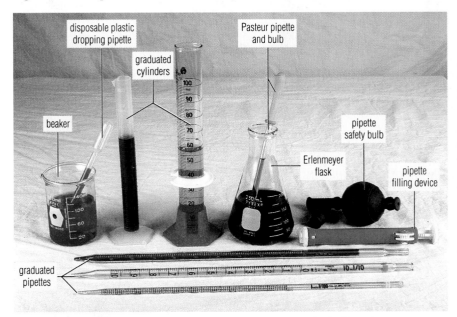

Figure 129c   Laboratory apparatus commonly used to contain liquids and to measure their **volume**. (Photo by J. W. Perry and D. Morton)

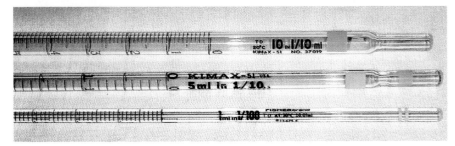

Figure 129d   Graduated pipettes. The graduations of the top two represent 1/10 (0.1) ml. The total volume measured with the top pipette is 10 ml, the middle one 5 ml, and the bottom one 1 ml. The graduations in the bottom pipette represent 1/100 (0.01) ml. (Photo by J. W. Perry and D. Morton)

Figure 129e   Reading the volume in a graduated cylinder. The correct reading is 42 ml. (Photo by J. W. Perry)

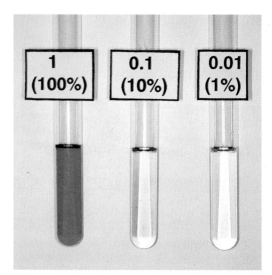

Figure 130a   Serial dilution. The concentration of the solution in the tube on the left has been reduced by a factor of 10 in the middle tube and by a factor of 100 in the tube on the right. (Photo by J. W. Perry)

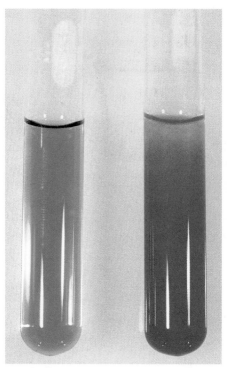

Figure 130b   Benedict's test for reducing sugars. The tube on the left has Benedict's solution added to distilled water. The tube on the right contains a reducing sugar (such as glucose) to which Benedict's solution has been added and then heated, illustrating a positive test for reducing sugar. (Photo by J. W. Perry)

Figure 130c   Potassium iodide (I₂KI) test for starch. The tube on the left has distilled water to which I$_2$KI has been added. The tube on the right contains a starch solution to which I$_2$KI has been added, illustrating a positive test for starch. (Photo by J. W. Perry)

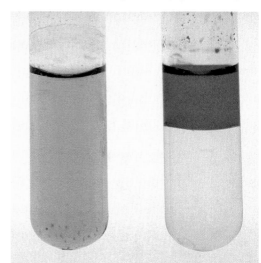

Figure 130d   Sudan test for lipids. The tube on the left has distilled water to which Sudan IV has been added. The tube on the right contains distilled water and a lipid to which Sudan IV has been added. Note that the lipid stains deep red and floats on top of the distilled water. (Photo by J. W. Perry)

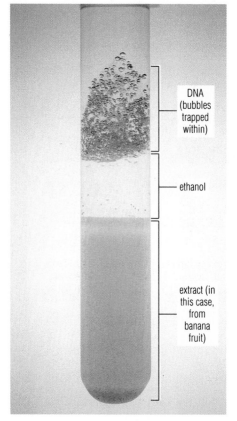

DNA (bubbles trapped within)

ethanol

extract (in this case, from banana fruit)

Figure 130e   DNA extraction. (Photo by J. W. Perry)

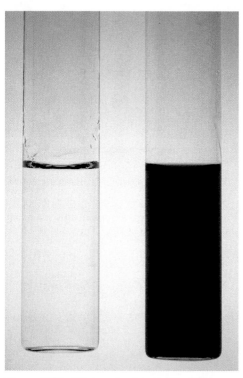

Figure 130f   Diphenylamine test for DNA. The tube on the left has distilled water to which diphenylamine has been added. The tube on the right contains DNA to which diphenylamine has been added and then heated, illustrating a positive test for DNA. (Photo by J. W. Perry)

Genera Index